Fabrizio Colombo • Irene Sabadini •
Daniele C. Struppa

Michele Sce's Works
in Hypercomplex Analysis

A Translation with Commentaries

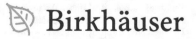 Birkhäuser

Fabrizio Colombo
Dipartimento di Matematica
Politecnico di Milano
Milano, Italy

Irene Sabadini
Dipartimento di Matematica
Politecnico di Milano
Milano, Italy

Daniele C. Struppa
Donald Bren Presidential Chair in Mathematics
Chapman University
Orange, CA, USA

ISBN 978-3-030-50218-8   ISBN 978-3-030-50216-4   (eBook)
https://doi.org/10.1007/978-3-030-50216-4

Mathematics Subject Classification: 30G35, 15A66, 15A78

This book is published under the imprint Birkhäuser, www.birkhauser-science.com, by the registered company Springer Nature Switzerland AG.
The registered company address is: Gewerbestrasse 11, 6330 Cham, Switzerland

Contents

Chapter 1
Introduction

1.1 Foreword

The richness of the theory of functions in one complex variable stimulated, at the end of the nineteenth century and beginning of twentieth century, the interest in the study of functions of several complex variables but also of hypercomplex variables. Some references about these studies can be found in Dickson's book [29], p. 78 and in the later paper by Ketchum [31]. Some earlier references are [27, 28].

In this context, the Italian school played a significant role in the twenties and thirties and in fact there are several works which range from the study of algebras of hypercomplex numbers, see e.g. the works of Scorza [35–39], to the analysis of functions of hypercomplex variables, see [40, 42, 43]. When the theory of monogenic (also called analytic or regular) functions of a quaternionic variable emerged with the works of Moisil [33] and Fueter [30] it was rather clear that the setting of quaternions, being a skew field, was a convenient one. The importance of these functions is witnessed by the fact that they appear in the monograph written by Segre [41]. But at that time, it seemed that there were different possible approaches, see for example the work of Sobrero [42] where a specific algebra is introduced to deal with equations arising from the theory of elasticity. These studies continued, see e.g. [44], and in the fifties attracted the attention of mathematicians like Rizza and Sce. Rizza was more interested in the case of Clifford algebras, see [34], whereas Sce was working in the context of more general algebras with a contribution to the octonionic case, together with Dentoni.

Most of the works of the Italian school have been essentially forgotten. It is difficult to understand the reason, but certainly one possible explanation is that many articles were published in Italian journals and they were written in Italian. Even today, and despite the existence of powerful search engines, it would be difficult to retrieve these papers, since the keywords one may use in the search are English words, not Italian ones.

F. Colombo et al., *Michele Sce's Works in Hypercomplex Analysis*,
https://doi.org/10.1007/978-3-030-50216-4_1

On the other hand, it is remarkable to note that Sce, in his works in hypercomplex algebras and analysis was very accurate about the existing literature and seemed to be aware of all relevant published works. As the reader will discover in his biography, he was a passionate bibliophile and this passion probably lead him to check very carefully the available references, most of them in foreign languages like English, German, Romanian.

The inspiration for this volume was given by the event "A Scientific Day in Honor of Michele Sce" held in Milano on October 11th, 2018. Michele Sce was professor for long time at the University of Milano where the three of us have studied. His interests in hypercomplex algebras and analysis constitute the common ground with our own research interests. The fact that the only results of Sce that are quoted in the modern literature were the one related to Fueter mapping theorem and, more marginally, the paper with Dentoni on octonions, has stimulated our willingness to translate his works in hypercomplex analysis to make them accessible. In doing so, we added some comments since sometimes it is necessary to adapt the notation and the terminology to the modern language, as well as some examples to clarify the results of Sce. Where appropriate, we provided the developments of the theory thus offering a deeper sense of the importance of Sce's work, and the visionary role he played in the theory of functions of hypercomplex variables.

We kept the original text and formulas of the various papers, and where we corrected some typos we point this out in the Editors' Notes. For typographical reasons, in Chap. 6 we used the modern Latex for Definitions, Theorems, etc. combined with the style used by the authors D_1, T_1, etc.

The book is organized in five chapters, besides this Introduction. Chapter 2 contains the translation of the three parts of a same paper dealing with the notions of monogenicity and total derivability in real and complex algebras. Chapter 3 deals with Sce's paper on systems of partial differential equations related to real algebras. Chapter 4 contains the paper on the variety of zero divisors in algebras. A central role is played by Chap. 5 which contains the celebrated theorem nowadays referred to as the Fueter-Sce-Qian mapping theorem which gave rise to several modern results discussed in the comments to this chapter. Chapter 6 contains the translation of the paper by Dentoni and Sce which deals with octonionic analysis, another topic which has led to many interesting modern developments. Each chapter has its own list of references and may stand alone.

Our hope is that this book will make Sce's work accessible to a larger audience and, possibly, will provide inspiration for future works.

1.2 Biography

Michele Sce was born in 1929 in Tirano in Northern Italy and he graduated "cum laude" at the Scuola Normale Superiore in Pisa in 1951. He became assistant professor in Geometry at the University of Milano where he remained officially until 1963, but already on leave at the end of 1962. During this period he lectured on function theory at the University of Parma, and on number theory at the University of Rome. He was the recipient, in 1959, of the Bonavera Prize awarded by the Academy of Sciences of Torino.

During this period Sce, among his many interests, was working on finite geometries and the enumerative problems connected to them but, when he tried to build examples or significant counterexamples, he faced difficult calculations. Thus he sought the help of his friend Lorenzo Lunelli who was among the first in Italy to work with an electronic processor at the Politecnico di Milano and had the "machine", namely the electronic calculator CRC-102A/P, capable of performing such calculations. The computations led them to the paper [2].

The CRC-102A/P, a machine with reduced computing capacity compared to the mathematical problems that can be resolved or at least clarified by automatic procedures, assisted Lunelli and Sce in writing four other papers. Given these interests, it is not surprising that Sce, at the end of 1962, left the university to take the opportunity to work at the Laboratorio di Ricerche Elettroniche of Olivetti. Sce then worked as a mathematical consultant at the Office of Electronic-Mechanical Equipment Projects directed by Pier Giorgio Perotto which was carrying out highly innovative projects such as the Olivetti Programma 101, the first personal computer in history, all designed in Italy.

Sce worked on projects concerning character recognition, working on projects on OCR-B (OCR stands for Optical Character Recognition), see [23], for which he was using a machine, the Elea 9003, to assist the computations.

Meanwhile, he married Paola Maria Manacorda and they eventually had three sons: Giovanni, Simone and Jacopo.

The Electronic Division of Olivetti was sold to General Electric in 1965 and Sce underlined in a newspaper that doing so Italy was losing a possible supremacy in the newly born computer industry. He continued to work for Olivetti as Head of the Research and Development Division, and he was still working for Olivetti when he was called to be a member of the Committee of Mathematics of the CNR (Consiglio Nazionale delle Ricerche) for the period 1968–1972. There he had the opportunity to illustrate his point of view on the state and the possible development of applied mathematics in Italy. He promoted scientific computing, also facilitating the creation of laboratories equipped with computers. He also envisioned and promoted the computer assisted teaching, an idea which was absolutely new and revolutionary at that time.

During the period he spent at Olivetti Sce was continuing his research and his teaching activity at the university. After he left Olivetti in 1971, he had various roles at the CNR where he was tasked to provide an impulse to the diffusion of computers in some Italian universities and to the preparation of curricula in computer science.

The next phase of his career led him to become Director of the Statistical Division of A.C. Nielsen in Milano, where he proposed a strategic plan based on data analysis.

He finally returned to academia in 1976 as Full Professor first at the University of Lecce, then in Torino and eventually, in 1980, back to the University of Milano where he taught several courses. As a teacher, he was available and very helpful to students, most of which still remember him for these qualities.

He had a strong commitment to mathematical libraries, initiating innovations in their management, and he took care of the Italian version of the Universal Decimal Classification of Mathematical Sciences on behalf of the CNR. He promoted the automation of Italian libraries for classification and management. In particular when he was Director of the library of the University of Milano, he adopted the system Aleph.

He also devoted himself to the dissemination of mathematics, both collaborating with magazines in the field and curating the edition of a large Dictionary of Mathematics published in 1989 by Rizzoli.

Sce has had many interests, a vast culture, and a personal library of about ten thousands volumes. This collection started when he was young and contains not only scientific books but also literary, historical, philosophical, anthropological books, as well as science fiction works and comics. A conspicuous part of the scientific books has been donated to the University of Milano Bicocca.

Despite his mild, shy and reserved nature, Sce always showed a great willingness to work with others in projects that aimed to strengthen the role of mathematics in Italian culture and society. As his wife recalled in her recollection he had few friends in academy, because of his reserved personality. Among them the late Carlo Pucci, Edoardo Vesentini and Gianfranco Capriz, and in Milano Stefano Kasangian, Stefania De Stefano and Alberto Marini. At Olivetti his best friends were Filippo Demonte and Mario Prennushi, with whom Sce collaborated while working at the project for the Programma 101.

Michele Sce passed away in Milano in 1993.

His ability to collaborate with various colleagues, his participation to numerous research projects and his dedication to teaching, left a trace of esteem and affection in his students, colleagues and friends, evidenced by the Scientific Day in his honor that the University of Milano organized on the 11th October 2018.

During that celebration it was particularly impressive to realize Sce's humble attitude. Despite his many achievements and his bright way of thinking and envisioning the future, he has never shown off his work with his family and friends. His various activities are now collected in the website www.michelesce.net which is a form of acknowledgement of his vision and understanding.

In the list of references below, the items [1–26] correspond to Michele Sce's scientific production.

Acknowledgments We are grateful to Professor Andrea D'Agnolo, editor in Chief of the *Rendiconti del Seminario Matematico della Università di Padova* and to Professor Carlo Sbordone, Editor in Chief of the *Atti della Accademia Nazionale dei Lincei* for giving the permission to translate the papers in the next chapters. We are grateful to Professor Marco Rigoli for the invitation to give a lecture at the event "A Scientific Day in Honor of Michele Sce", which gave us the idea for this project.

The authors are indebted to Paula Cerejeiras, Antonino De Martino, Kamal Diki, Elena Luna Elizarraras, Uwe Kähler, Andrea Previtali, Michael Shapiro for their useful comments and for pointing out some further references.

Many thanks are due to Giuliano Moreschi of the Library of the Università degli Studi di Milano for giving access to the authors to some old books, and to Sonia Pasqualin of the Library of the Politecnico di Milano, for her assistance in finding several old papers.

The authors are also grateful to Paola Maria Manacorda, Michele Sce's wife, for sharing some memories and to Giovanni, Simone and Jacopo Sce, their sons, for the material which was the source for the biography in the next section.

References

1. Dentoni, P., Sce, M.: Funzioni regolari nell'algebra di Cayley. Rend. Sem. Mat. Univ. Padova **50**, 251–267 (1973)
2. Lunelli, L., Sce, M.: Sulla ricerca dei k-archi completi mediante una calcolatrice elettronica. In: UMI (ed.) Convegno Internazionale: Reticoli e Geometrie Proiettive, Palermo, 25–29 Ottobre 1957 Messina 30 Ottobre 1957 (Ed. Cremonese Roma, 1958), pp. 81–86
3. Lunelli, L., Sce, M.: K-archi completi nei piani proiettivi desarguesiani di rango 8e16. Centro di calcoli numerici, Politecnico di Milano (1958)
4. Lunelli, L., Sce, M.: Tavola dei K_{11}-archi completi per cinque punti. Rend. Ist. Lombardo Accad. Sci. Lett. Cl. Sci. (A) **95**, 92–102 (1961)
5. Lunelli, L., Sce, M.: Considerazioni aritmetiche e risultati sperimentali sui $\{K; n\}_q$-archi. Rend. Ist. Lombardo Accad. Sci. Lett. Cl. Sci. (A) **98**, 3–52 (1964)
6. Lunelli, L., Sce, M., Lunelli, M.: Calcolo di omografie cicliche di piani desarguesiani finiti. Rend. di Mat. e sue Appl. Serie V **18**, 351–374 (1959)
7. Sce, M.: Su alcune proprietà delle matrici permutabili e diagonizzabili. Rivista Mat. Univ. Parma **1**, 363–374 (1950)
8. Sce, M.: Osservazioni sulle forme quasi-canonica e pseudo-canonica delle matrici. Rend. Sem. Mat. Univ. Padova **19**, 324–339 (1950)
9. Sce, M.: Su una generalizzazione delle matrici di Riemann: memoria prima. Ann. Scuola Norm. Super. Pisa (3) **5**, 81–103 (1951)
10. Sce, M.: Su una generalizzazione delle matrici di Riemann: memoria seconda. Ann. Scuola Norm. Super. Pisa (3) **5**, 301–327 (1951)
11. Sce, M.: Sugli r birapporti di $r + 3$ punti di un S_r. Rend. Ist. Lomb. Accad. Sci. Lett. Cl. Sci. Mat. Nat. (3) **16**(85), 363–374 (1952)
12. Sce, M.: Monogeneità e totale derivabilità nelle algebre reali e complesse, Nota I. Atti Accad. Naz. Lincei. Rend. Cl. Sci. Fis. Mat. Nat. (8) **16**, 30–35 (1954)
13. Sce, M.: Monogeneità e totale derivabilità nelle algebre reali e complesse, Nota II. Atti Accad. Naz. Lincei. Rend. Cl. Sci. Fis. Mat. Nat. (8) **16**, 188–193 (1954)
14. Sce, M.: Monogeneità e totale derivabilità nelle algebre reali e complesse, Nota III. Atti Accad. Naz. Lincei. Rend. Cl. Sci. Fis. Mat. Nat. (8) **16**, 321–325 (1954)
15. Sce, M.: Sui sistemi di equazioni a derivate parziali inerenti alle algebre reali. Atti Accad. Naz. Lincei. Rend. Cl. Sci. Fis. Mat. Nat. (8) **18**, 32–38 (1955)
16. Sce, M.: Sul postulato della continuità della retta ed alcune sue applicazioni alla geometria elementare. Period. Mat. (4) **33**, 215–225 (1955)
17. Sce, M.: Sul postulato della continuità della retta ed alcune sue applicazioni alla geometria proiettiva. Period. Mat. (4) **33**, 297–308 (1955)
18. Sce, M.: Sulla varietdei divisori dello zero nelle algebre. Atti Accad. Naz. Lincei. Rend. Cl. Sci. Fis. Mat. Nat. (8) **23**, 39–44 (1957)
19. Sce, M.: Osservazioni sulle serie di potenze nei moduli quadratici. Atti Accad. Naz. Lincei. Rend. Cl. Sci. Fis. Mat. Nat. (8) **23**, 220–225 (1957)
20. Sce, M.: Sui k_q-archi di indice h. In: UMI (ed.) Convegno Internazionale: Reticoli e Geometrie Proiettive, Palermo, 25–29 Ottobre 1957 Messina 30 Ottobre 1957 (Ed. Cremonese Roma, 1958), pp. 133–135

21. Sce, M.: Sulla completezza degli archi nei piani proiettivi finiti: nota 1. Atti Accad. Naz. Lincei. Rend. Cl. Sci. Fis. Mat. Nat. (8) **25**, 43–51 (1958)
22. Sce, M.: Preliminari ad una teoria aritmetico gruppale dei k-archi. Rend. Mat. e Appl. Serie V **19**(3/4), 241–291 (1960)
23. Sce, M.: Riconoscimento di caratteri. Calcolo **1**, 267–298 (1964)
24. Sce, M.: Sugli sviluppi di Charlier. In: Pubblicazioni dell'Istituto per le Applicazioni del Calcolo "Mauro Picone" (IAC), Serie III, 200 Ed., 54 pp. Marves, Roma (1979)
25. Sce, M.: Polinomi ortogonali, successioni fattoriali e momenti. Rend. Sem. Mat. Univ. Politec. Torino **39**(2), 61–97 (1981/1982)
26. Sce, M.: Geometria combinatoria e geometrie finite. Rend. Sem. Mat. Fis. Milano **51**(1981), 77–123 (1983)

List of References for this Introduction

27. Autonne, L.: Leipziger Berichte **45**, 828 (1893)
28. Autonne, L.: Sur la fonction monogène d'une variable hypercomplexe dans un groupe commutatif. Bull. Soc. Math. France **37**, 176–196 (1909)
29. Dickson, L.E.: Linear Algebras. Cambridge University, Cambridge (1914)
30. Fueter, R.: Analytische Funktionen einer Quaternionenvariablen. Comment. Math. Helv. **4**, 9–20 (1932)
31. Ketchum, P.W.: Analytic functions of hypercomplex variables. Trans. Am. Math. Soc. **30**(4), 641–667 (1928)
32. Kriszten, A.: Hypercomplexe und pseudo-analytische Funktionen. Comm. Math. Helv. **26**, 6–35 (1952)
33. Moisil, G.C.: Sur les quaternions monogènes. Bull. Sci. Math. (Paris) **LV**, 168–174 (1931)
34. Rizza, G.B.: Sulle funzioni analitiche nelle algebre ipercomplesse. Pont. Acad. Sci. Comment. **14**, 169–194 (1950)
35. Scorza, G.: Le algebre doppie. Rend. Acc. Napoli (3) **28**, 65–79 (1922)
36. Scorza, G.: Sopra un teorema fondamentale della teoria delle algebre. Rend. Acc. Lincei (6) **20**, 65–72 (1934)
37. Scorza, G.: Le algebre per ognuna delle quali la sottoalgebra eccezionale è potenziale. Atti R. Acc. Scienze di Torino **70**(11), 26–45 (1934–1935)
38. Scorza, G.: Le algebre del 3° ordine. Rend. Acc. Napoli (2) **20**, n.13 (1935)
39. Scorza, G.: Le algebre del 4° ordine. Rend. Acc. Napoli (2) **20**, n.14 (1935)
40. Scorza Dragoni, G.: Sulle funzioni olomorfe di una variabile bicomplessa. Mem. Acc. d'Italia **5**, 597–665 (1934)
41. Segre, B.: Forme differenziali e loro integrali. Roma (1951)
42. Sobrero, L.: Algebra delle funzioni ipercomplesse e una sua applicazione alla teoria matematica dell'elasticità. Mem. R. Acc. d'Italia **6**, 1–64 (1935)
43. Spampinato, N.: Sulle funzioni totalmente derivabili in un'algebra reale o complessa dotata di modulo. Rend. Lincei **21**, 621–625 (1935)
44. Ward, J.A.: A theory of analytic functions in linear associative algebras. Duke Math J. **7**, 233–248 (1940)

Chapter 2
Monogenicity and Total Derivability in Real and Complex Algebras

In this chapter we collect three papers that correspond to the translations of three parts of the same work originally published as:

M. Sce, *Monogeneità e totale derivabilità nelle algebre reali e complesse. I*, (Italian) Atti Accad. Naz. Lincei. Rend. Cl. Sci. Fis. Mat. Nat. (8) **16** (1954), 30–35.

M. Sce, *Monogeneità e totale derivabilità nelle algebre reali e complesse. II*, (Italian) Atti Accad. Naz. Lincei. Rend. Cl. Sci. Fis. Mat. Nat. (8) **16** (1954), 188–193.

M. Sce, *Monogeneità e totale derivabilità nelle algebre reali e complesse. III*, (Italian) Atti Accad. Naz. Lincei. Rend. Cl. Sci. Fis. Mat. Nat. (8) **16** (1954), 321–325.

2.1 Monogenicity and Total Derivability in Real and Complex Algebras, I

Article I by Michele Sce, presented during the meeting of 16 January 1954 by B. Segre, member of the Academy.

To construct a theory of functions of a hypercomplex variable, a natural way would be to generalize the function theory of a complex variable. However, to pass from functions of a complex variable (for which the uniqueness of the derivative follows from the monogenicity condition) to functions of a hypercomplex variable, there are two possibilities: one is to impose the uniqueness of the derivative, and

© The Editor(s) (if applicable) and The Author(s), under exclusive licence to Springer Nature Switzerland AG 2020
F. Colombo et al., *Michele Sce's Works in Hypercomplex Analysis*, https://doi.org/10.1007/978-3-030-50216-4_2

this yields to the theory of *totally derivable functions*;[1] the second is to generalize the monogenicity conditions and this yields to the theory of *monogenic functions*.[2]

In the course held at the Istituto di Alta Matematica in the year 1952–1953, Prof. B. Segre proposed to study algebras for which the notion of total derivability implies the one of monogenicity.

This Note I deals with the search of the conditions that the basis of an algebra, for simplicity we assume any algebra with module [Editors' note: here module means unit. In the following, we will always translate the term *module* with the modern term *unit*], must satisfy in order for this to hold. The forthcoming Note II, III will deal with the case of algebras of order two, three, four,[3] and some cases of higher order algebras, whose bases satisfy these conditions. We believe that our results for algebras of the fourth order, in which case we find five algebras, are particularly interesting. Four of them (precisely the bicomplex, the bidual, the quaternions and the algebra of matrices of the second order) are already widely studied; we think that the fifth is considered here for the first time and it would maybe deserve a deeper study.

1. Let $\mathscr{A}$ be a real or complex algebra of order n, with unit. Given a basis $u = (u_1, \ldots, u_n)$ we say that the algebras $\mathscr{A}'$, $\mathscr{A}''$ are the *first and second regular representation*[4] of $\mathscr{A}$ if their elements are order n matrices X', X'' defined, for any $x \in \mathscr{A}$, by the relations

$$xu = uX' \tag{2.1}$$

$$ux = uX''_{-1}.\,^5 \tag{2.2}$$

---

[1] N. Spampinato, *Sulle funzioni totalmente derivabili in un'algebra reale o complessa dotata di modulo*, Rend. Lincei, vol. 21 (1935), I, 621–625, II, 683–687. Functions totally derivable over the bicomplex numbers have been studied by G. Scorza Dragoni, *Sulle funzioni olomorfe di una variabile bicomplessa*, Mem. Acc. d'Italia, vol. 5 (1934), 597–665; in the bidual algebra by L. Sobrero, *Algebra delle funzioni ipercomplesse e una sua applicazione alla teoria matematica dell'elasticità*, Mem. Acc. d'Italia, vol. 6 (1935), 1–64.

[2] This is the terminology used by B. Segre, *Forme differenziali e loro integrali*, Roma, 1951 and, in the particular case of quaternions by Gr. C. Moisil, *Sur les quaternions monogènes*, Bull. Sci. Math. (Paris), **LV** (1931), 168–174. R. Fueter uses the terminology regular functions, *Über die Funktionentheorie in einer hypercomplexen Algebra*, Elem. Math., III, 5 (1948), 89–94 and this is the term used by his school and by G. B. Rizza, *Sulle funzioni analitiche nelle algebre ipercomplesse*, Comm. Pont. Ac. Sc., vol. 14 (1950), 169–174. This last Author calls monogenic the totally derivable functions.

[3] Algebras of these orders on any field have been classified by G. Scorza in the works *Le algebre doppie*, Rend. Acc. Napoli (3), vol. 28 (1922), 65–79, *Le algebre del 3° ordine*, Acc. Napoli (2), vol. 20 (1935), n.13, *Le algebre del 4° ordine*, ibid. n. 14.

[4] Cfr. A. A. Albert, *Structure of algebras*, American Mathematical Society Colloquium Publications, Volume XXIV, New York, 1939.

[5] In the sequel we will always consider u and its transpose u_{-1} as n-dimensional vectors, the first as a row, the second as a column [Editors' note: the same notation is used for matrices].

We say that an element

$$y = y_1 u_1 + \cdots + y_n u_n = (y_1, \ldots, y_n)(u_1, \ldots, u_n)_{-1} = \eta u_{-1}$$

belonging to $\mathscr{A}$ is a *right or left totally derivable function* of an element

$$x = x_1 u_1 + \cdots + x_n u_n = \xi u_{-1}$$

in $\mathscr{A}$, if the jacobian matrix $\partial \eta / \partial \xi = dy/dx$ belongs to $\mathscr{A}'$ or its transpose belongs to $\mathscr{A}''$.[6]

Finally, we say that the element y in $\mathscr{A}$ is a *right or left monogenic function*[7] of an element x in $\mathscr{A}$, if

$$u \frac{dy}{dx} u_{-1} = 0 \tag{2.3}$$

or

$$u \left(\frac{dy}{dx} \right)_{-1} u_{-1} = 0. \tag{2.4}$$

When performing the change of basis

$$u' = u P^{-1}, \tag{2.5}$$

since

$$x = \xi u_{-1} = \xi' u'_{-1}, \qquad y = \eta u_{-1} = \eta' u'_{-1}$$

[Editors' note: it was Y in the original manuscript] it turns out that

$$\xi' = \xi P_{-1}, \qquad \eta' = \eta P_{-1}$$

and so

$$\frac{d\eta}{d\xi} = \frac{d\eta}{d\eta'} \frac{d\eta'}{d\xi'} \frac{d\xi'}{d\xi} = P^{-1} \frac{d\eta'}{d\xi'} P.$$

Thus *the definition of totally derivable functions is invariant with respect to change of basis* (indeed

$$xu' = xu P^{-1} = uX'P^{-1} = u'PX'P^{-1} = u'X'^*$$

---

[6]See N. Spampinato cited in [(1)].
[7]See B. Segre cited in [(2)], p. 442.

with $X' = P^{-1}X'^*P$) when performing the change of basis (2.5) in fact formulas (2.3) and (2.4) become

$$u'\frac{\partial \eta'}{\partial \xi'}PP_{-1}u'_{-1} = 0 \tag{2.6}$$

$$u'PP_{-1}\left(\frac{\partial \eta'}{\partial \xi'}\right)_{-1} u'_{-1} = 0; \tag{2.7}$$

the conditions (2.6) *and* (2.7) *depend, in general, on the change of basis,*[8] in the sense that they ensure the existence (when $|P| \neq 0$) of a suitable basis $u = u'P$ such that in that basis y is a monogenic function of x.

After that, the problem of comparing the notion of monogenic function with the one of totally derivable function translates into the search of conditions under which a totally derivable function is monogenic *with respect to a suitable basis*, namely in the comparison between the conditions that dy/dx belongs to $\mathscr{A}'$ and formulas (2.6) and (2.7).

2. We now consider the functions

$$y_i(x) = u_i x = u\xi'_{-1}, \qquad i = 1, \ldots, n;$$

by virtue of (2.1), we can write

$$u_i x = u_i u\xi_{-1} = uU'_i\xi_{-1},$$

with U'_i in $\mathscr{A}'$, and it turns out that

$$\xi'_{-1} = U'_i\xi_{-1}.$$

But then

$$\frac{du_i x}{dx} = U'_i$$

and $y_i(x) = u_i x$ *are right totally derivable.*

Thus, if in our algebra the right totally derivable functions are also monogenic, there should exists a basis such that $u_i x$ are right or left monogenic in x, that is, it should hold

$$uU'_i PP_{-1}u_{-1} = 0, \tag{2.8}$$

---

[8]The conditions coincide with (2.3) when PP_{-1} is a scalar matrix; however, if the matrices in $\mathscr{A}'$ are direct sum of matrices then the equality with (2.3) also for suitable diagonal, not scalar matrices PP_{-1}. This property of the elements in $\mathscr{A}'$ is possessed by the decomposable algebras, but also by indecomposable algebras like the one of ternions with the basis given in n. **4** [Editors' note: it was n. **5** in the original manuscript] but this can be overcome by considering the matrix P.

or

$$u P P_{-1}(U_i')_{-1} u_{-1} = 0, \tag{2.9}$$

for some nonsingular matrix P.[9]

Since U_i' are elements in $\mathscr{A}'$, (2.1), (2.8), and (2.9) allow to deduce

$$u_i u P P_{-1} u_{-1} = 0, \qquad i = 1, \ldots, n \tag{2.10}$$

$$u P P_{-1} u_i u_{-1} = 0, \qquad i = 1, \ldots, n. \tag{2.11}$$

As each element z of $\mathscr{A}$ is a linear combination (with real or complex coefficients) of u_i, by taking a linear combination of (2.10) and (2.11) one has

$$z u P P_{-1} u_{-1} = 0, \tag{2.12}$$

or

$$u P P_{-1} z u_{-1} = 0; \tag{2.13}$$

conversely, if (2.12) and (2.13) hold for each element z in $\mathscr{A}$, then they hold also for u, thus one reobtains (2.10) and (2.11).

In particular, when taking z equal to the unit then (2.12) and (2.13) give

$$u P P_{-1} u_{-1} = 0; \tag{2.14}$$

from this one reobtains in an obvious way both (2.12) and (2.10), so these latter are equivalent to (2.14).[10]

Given a right totally derivable function $y(x)$, its jacobian matrix dy/dx will automatically belong to $\mathscr{A}'$; let z be the corresponding element in $\mathscr{A}$ and let us assume that (2.14) or (2.11) hold. Then also (2.12) or (2.13) hold, so that because of (2.1), we reobtain (2.6) or (2.7); thus $y(x)$ is right or left monogenic.

Thus we may conclude that (2.14) *and* (2.11) *with P nonsingular, are necessary and sufficient conditions for right totally derivable functions in an algebra $\mathscr{A}$ to be right or left monogenic (with respect to a suitable basis); condition* (2.14) *is necessary also for the left monogenicity.* [Editors' Note: see Remark 2.4][11]

---

[9]Evidently the difference between right and left monogenic is not relevant in commutative algebras.

[10]Since, by virtue of (2.1), from (2.10) it follows (2.8), then (2.14) is also equivalent to all equations (2.8).

[11]In an analogous way one can prove that (2.14) *and* (2.11) *are necessary and sufficient conditions for left totally derivable functions to be monogenic on the left or on the right.* [Editors' Note: see Remark 2.4]

This latter assertion ensures that it is necessary that $y(x)$ be right monogenic in order to have that right total derivability imply left monogenicity; thus *the algebras in which right total derivability implies left monogenicity are the algebras in which right totally derivable functions are both right and left monogenic.*

Let us now recall that the monogenicity conditions are a system of n differential equations while those of total derivability are a system of $n(n - 1)$ differential equations;[12] moreover the conditions for right and left monogenicity are $n + m \leq 2n$.

Since (2.14) or (2.11) translate into linear conditions on the basis elements, and thus in conditions concerning only the algebra, in order to get right total differentiability provided that (2.14) or (2.11) hold, one has to add $n(n - 2)$ differential equations to the n arising from right monogenicity or to the $n+m$ arising from both the right and left monogenicity.

3. We say that an algebra with unit is *solenoidal* if its bases satisfy relations of the form (2.14) with a nonsingular P and we say, in particular, that it is *bisolenoidal* if its bases satisfy relations of the form (2.11).

Given two algebras $\mathscr{A} = (u_1, \ldots, u_n)$, $\mathscr{B} = (v_1, \ldots, v_m)$ we consider their direct sum $\mathscr{S}$ and their direct product $\mathscr{P}$ whose basis are, respectively

$$w = (u_1^o, \ldots, u_n^o, v_1^o, \ldots, v_m^o) = (u^o; v^o)$$

$$w' = (u_1^o v_1^o, u_1^o v_2^o, \ldots, u_n^o v_m^o) = (u_1^o v^o; \ldots; u_n^o v^o),$$

where

$$\mathscr{A}_o = (u_1^o, \ldots, u_n^o), \qquad \mathscr{B}_o = (v_1^o, \ldots, v_m^o)$$

are algebras isomorphic to $\mathscr{A}, \mathscr{B}$.[13] If $\mathscr{A}$ and $\mathscr{B}$ are bisolenoidal, so are also $\mathscr{A}_o$ and $\mathscr{B}_o$ so there exist nonsingular matrices P and Q such that

$$\begin{aligned} u^o P P_{-1} u_i^o u_{-1}^o = 0, \qquad & i = 1, 2, \ldots, n, \\ v^o Q Q_{-1} v_k^o v_{-1}^o = 0, \qquad & k = 1, 2, \ldots, m. \end{aligned} \tag{2.15}$$

From this, by setting

$$R = \begin{pmatrix} P & 0 \\ 0 & Q \end{pmatrix}$$

[12] See Segre cited in [2], p. 443 and p. 451.

[13] See A. A. Albert, *Modern higher Algebra*, The University of Chicago Science Series, Chicago, 1937, Chap. X, n. 4 and Albert, cited in [4], Chap. 1, n.5.

and recalling that in the direct sum $u_i^o v_k^o = v_k^o u_i^o = 0$, one gets

$$w R R_{-1} u_i^o w_{-1} = w R R_{-1} \begin{pmatrix} u_i u_{-1}^o \\ 0 \end{pmatrix}$$

$$= (u^o \ \ v^o) \begin{pmatrix} P P_{-1} & 0 \\ 0 & Q Q_{-1} \end{pmatrix} \begin{pmatrix} u_i u_{-1}^o \\ 0 \end{pmatrix}$$

$$= u^o P P_{-1} u_i^o u_{-1}^o = 0;$$

and, analogously, one finds that

$$w R R_{-1} v_k^o w_{-1} = 0.$$

Thus, if w_i is any element in w, there exists a nonsingular matrix R such that

$$w R R_{-1} w_i w_{-1} = 0,$$

i.e., *the direct sum of bisolenoidal algebras is bisolenoidal.* Assume that only $\mathscr{B}$ is solenoidal, that is, (2.15) holds; then, by setting

$$R = Q \times I_n = \begin{pmatrix} Q & \dots & 0 \\ \vdots & \ddots & \vdots \\ 0 & \dots & Q \end{pmatrix}$$

and recalling that in the direct product $u_i v_k = v_k u_i$, one obtains

$$w' R R_{-1} u_i^o v_k^o w_{-1}' = w' R R_{-1} v_k^o \begin{pmatrix} v_{-1}^o u_i^o u_1^o \\ \vdots \\ v_{-1}^o u_i^o u_n^o \end{pmatrix}$$

$$= (u_1^o v^o, \dots, u_n^o v^o) \begin{pmatrix} Q Q_{-1} & \dots & 0 \\ \vdots & \ddots & \vdots \\ 0 & \dots & Q Q_{-1} \end{pmatrix} \begin{pmatrix} v_k^o v_{-1}^o u_i^o u_1^o \\ \vdots \\ v_k^o v_{-1}^o u_i^o u_n^o \end{pmatrix}$$

$$= u_1^o v^o Q Q_{-1} v_k^o v_{-1}^o u_i^o u_1^o + \dots + u_n^o v^o Q Q_{-1} v_k^o v_{-1}^o u_i^o u_n^o = 0.$$

Thus, *the direct product of algebras, one of which is bisolenoidal is bisolenoidal.*[14]

---

[14]Obviously, an algebra which is direct product may be bisolenoidal even when one of the factors is not.

2.2 Monogenicity and Total Derivability in Real and Complex Algebras, II

Article II by Michele Sce, continuation of Article I published at p. 30 of this volume, presented during the meeting of 13 February 1954 by B. Segre, member of the Academy.

4. Setting $PP_{-1} = \|a_{ik}\|$, $(i, k = 1, \ldots, n)$, formula (2.14) rewrites as $\sum_{i,k} a_{ik} u_i u_k = 0$ [Editors' note: $\|a_{ik}\|$ denotes the matrix with elements a_{ik}]; thus, since the units u_1 and u_2 of the complex algebras of the second order, i.e., of the complex and dual numbers,[15] combine according to the rules $u_1 u_i = u_i$, $u_2^2 = -u_1$, and $u_1 u_i = u_i$, $u_2^2 = 0$, $(i = 1, 2)$, it turns out that to satisfy (2.14), it must be $a_{11} = a_{22}$, $a_{12} = 0$ or $a_{11} = a_{12} = 0$. In the second case, PP_{-1} is singular and so *the algebra of dual numbers is not solenoidal and the algebra of complex numbers is the only complex algebra solenoidal of the second order.* We now try to satisfy (2.14) for two of the five complex algebras of the third order, [16] tripotential and tridual numbers, whose units combine according to the rules $u_1 u_i = u_i$, $u_2^2 = u_3$, and $u_1 u_i = u_i$, $(i = 1, 2, 3)$; in the first case one has $a_{11} = a_{22} = 0$ and $a_{13} + a_{22} = 0$, in the second case $a_{1i} = 0$ and PP_{-1} is singular. An analogous analysis for the remaining three algebras shows that *only the algebra of tridual numbers is not solenoidal.*

In the case of ternions whose units combine according to the rules $u_1^2 = u_1$, $u_2^2 = u_2$, and $u_1 u_3 = u_3 u_2 = u_3$, (2.11) translate into

$$a_{11} u_1 + a_{13} u_3 = 0, \quad a_{22} u_2 + a_{32} u_3 = 0, \quad a_{12} u_3 = 0 \tag{2.16}$$

which ensure that PP_{-1} is singular; thus *the algebra of ternions is not bisolenoidal,*[17] *and among complex algebras of the third order, only the solenoidal commutative algebras are bisolenoidal.* Considering the 16 multiplication tables which arise from the complex algebras of the 4th order[18] together with (2.14) it can be proved that *only the algebras with multiplication tables XLI and LV*

---

[15] See G. Scorza, first work cited in I, (3). It is understood that we always consider algebras with unit, and different up to isomorphisms.

[16] See G. Scorza, second work cited in I, (3). In the classification in [32] the five algebras have the multiplication tables III (tricomplex numbers), X (tripotential numbers), XI (tridual numbers), XXIV (direct sum of dual numbers and the complex field) and XXVII (ternions). Ternions are the only noncommutative algebra. In the text, instead of tables, we will make use of the multiplication rules in which the vanishing products $u_i u_k$ will not appear.

[17] Since the unit of the algebra is $u_1 + u_2$, (2.14) can be obtained by summing the first and the second relation in (2.16) leading to $a_{11} = a_{22} = a_{13} + a_{32} = 0$; thus with the change of basis $u_i' = u_i + u_2$, $u_2' = i(u_1 - u_2)$, $u_3' = u_3$, one gets a basis such that $u'u'_{-1} = 0$.

[18] See G. Scorza, third work cited in I, (3). ; in this classification the 16 tables are I* (quaternions), X (quadricomplex numbers), XXXVI (quadripotential numbers), XXXVII* (∞^1 many non isomorphic algebras depending on a parameter), XXXIX (two algebras for the values $0, 1$ of a parameter), XLI*, LV (quadridual numbers), XC, C*, CIII*, CIV*, CV, CVIII*, CXXV, CXXVIII*. The asterisk denotes the noncommutative algebras.

are not solenoidal. To this end, we limit ourselves to observe that since the multiplication rules in the two algebras are, respectively, $u_1 u_i = u_i$ and $u_1 u_i = u_i$, $u_2 u_3 = -u_3 u_2 = u_4$ $(i = 1, \ldots, 4)$, in both cases to satisfy (2.14) the first row of PP_{-1} must vanish.[19]

Imposing (2.11) for the six noncommutative algebras, one concludes that *the only bisolenoidal algebras are those whose units satisfy* $u_1 u_i = u_i$, $u_2^2 = u_2$, $u_3 = -u_3$, $u_2 = u_4$, $u_3^3 = \alpha u_4$.[20]

5. In the study of real solenoidal algebras, it is important to bear in mind that also the matrix P in (2.14) and (2.11) is real, thus PP_{-1} is symmetric and positive definite.[21] . This remark allows to have the converse of the first Theorem in n. 3, namely to show that *a direct sum of real algebras is bisolenoidal only if its components are bisolenoidal.*

Indeed, if the real algebra $\mathscr{C} = (w) = (u^o \, v^o)$, direct sum of the algebras $\mathscr{A}$ and $\mathscr{B}$, is bisolenoidal there exists a symmetric, positive definite matrix

$$A = \begin{pmatrix} A_1 & A_2 \\ A_3 & A_4 \end{pmatrix}$$

such that

$$w A w_i w_{-1} = u^o A_1 w_i u^o_{-1} + v^o A_4 w_i v^o_{-1} = 0;$$

from this relation, and according to the fact that w_i is an element either in u^o or in v^o, one gets

$$u^o A_1 u_i^o u^o_{-1} = 0, \qquad v^o A_4 u_i^o u^o_{-1} = 0$$

with A_1 and A_4 still symmetric and positive definite since they are principal minors of A.

It turns out that *the only solenoidal algebra of order two is the one of complex numbers.* In fact, besides the two algebras over the complex field, there is the algebra of bireal numbers[22] which is direct sum of the real field (certainly non solenoidal) with itself.

---

[19]This observation extends to the n-dual numbers whose multiplication rules are $u_1 u_i = u_i$ $(i = 1, \ldots, n)$ and it can be proved that *the algebras of n-dual numbers are not bisolenoidal*

[20]These multiplication rules translate into Scorza's table XXXVII after the change of basis $u_1' = u_1$, $u_2' = u_2$, $u_3' = \frac{1}{2}(u_2 - u_3)$, $u_4' = u_4$. We point out that with respect to the basis $u_1' = u_1 - u_4$, $u_2' = u_2 + u_3$, $u_3' = iu_1$, $u_4' = u_2 + iu_3$, (2.14) becomes $u' u'_{-1} = 0$ which has as immediate consequence the four relations in (2.11).

[21]See A. A. Albert, *Modern higher algebra*, (Chicago, 1937), Chap. V, n.12.

[22]See Scorza cited in [1].

About the six algebras of the third order,[23] we have already seen in n. 4 that the three indecomposable algebras cannot satisfy (2.14) with PP_{-1} symmetric and positive definite; since the three decomposable algebras are not solenoidal by the theorem just proven, one can conclude that *there are no real algebras solenoidal of the third order*.

Among the real algebras of the fourth order[24] *in addition to the two algebras direct product of the algebra of complex numbers with the one of the bireal numbers and of the dual numbers, only the algebra of quaternions, the algebra of 2×2 matrices and the one with multiplication table LXXXI are solenoidal; none of these algebras is bisolenoidal.*

Making use of a direct proof, or of theorems that we will provide in n. 6, one can prove that among the algebras of the fourth order only the five mentioned in the assertion can be solenoidal; since, by the second theorem in n. 3, the two direct product algebras are solenoidal, we shall examine only the three noncommutative algebras.

The units of the algebras in table LXXXI can be combined according to the multiplication rules

$$u_1 u_i = u_i, \quad (i = 1, \ldots, 4), \quad u_2 u_3 = -u_3 u_2 = u_4, \quad u_2 u_4 = -u_4 u_2 = -u_3,$$

[Editors' note: one needs also the condition $u_2^2 = -u_1$] thus (2.14) leads to

$$a_{11} - a_{22} = a_{12} = a_{13} = a_{14} = 0$$

which ensures the fact that the algebra is solenoidal;[25] adding $(a_{11} + a_{22})u_3 = 0$ to (2.14), we reobtain all (2.11) which, however, can be satisfied only with PP_{-1} singular and the algebra is not bisolenoidal.

For the algebra of 2×2 matrices, if we select the units $e_{i,k}$, $(i, k = 1, 2)$ which combine according to $e_{i,h}e_{h,k} = e_{i,k}$, relations (2.11) give rise to

$$\sum_i a_{j,h+i}e_{1,i} + \sum_i a_{j+2,h+i}e_{2,i} = 0,$$

$(i, j = 1, 2; h = 0, 2)$; since, for $h = 0$ they impose the vanishing of the first two columns of PP_{-1} and for $h = 2$ of the remaining two columns, the algebra is not bisolenoidal, not even in the complex field. As the unit of the algebra is $e_{11} + e_{22}$,

---

[23] See Scorza cited in [(2)]; besides the tables in [(2)], we have to add I (direct sum of the algebras of complex and real numbers) which, over the complex field, reduces to III. We recall that the classification given by Scorza is independent of the field of numbers.

[24] See Scorza cited in [(4)]; besides the tables listed therein, we have to add III* (2×2 matrices), V (direct product of the complex numbers and the bireal), VIII, LXXIX (direct product of complex numbers with the bidual), LXXXI* and CXIX which, in the complex case, may be reduced to I, X, X, CV, CVIII, CXXV, respectively.

[25] The fact that it is solenoidal is in fact evident, since $uu_{-1} = 0$.

(2.14) can be obtained by summing the relations that we have for $j = 1, h = 0$, and $j = h = 2$ and this leads to

$$a_{11} + a_{23} = a_{12} + a_{24} = a_{31} + a_{43} = a_{32} + a_{44} = 0;$$

these conditions are compatible with the fact that PP_{-1} is symmetric, positive definite so that the algebra is solenoidal.[26]

In the algebra of quaternions, whose basis is $e_0 = 1, e_1, e_2, e_3 = e_1e_2$ and satisfies

$$e_1^2 = e_2^2 = -1, \quad e_1e_2 + e_2e_1 = 0,$$

(2.14) rewrites as

$$a_{11} - \sum_{k=2,3,4} a_{kk} + 2 \sum_{k=2,3,4} a_{1k}e_{k-1} = 0;$$

thus one gets $a_{11} = \sum_{k=2,3,4} a_{kk}, a_{1k} = 0, (k = 2, 3, 4)$ and since these conditions are compatible with the fact that PP_{-1} is positive definite we get that the algebra is solenoidal.

Since $e_1(e_0, \ldots, e_3) = 2(e_1, -e_0, 0, 0) - (e_0, \ldots, e_3)e_1$, the second of (2.11), once that the first one of the (2.11) is satisfied namely (2.14), reduces to

$$(a_{11} - a_{22})e_1 - \sum_{k=3,4} a_{2k}e_{k-1} = 0,$$

and thus it leads to

$$\sum_k a_{kk} = 0, \quad a_{2k} = 0, \quad (k = 3, 4).$$

Then, imposing the remaining (2.11), one obtains that PP_{-1} must vanish and this excludes that the algebra is bisolenoidal, also over the complex field.

6. Real division algebras are, in addition to the field of real numbers, the algebra of complex numbers and the one of quaternions;[27] thus, from n. 5, we deduce that also real division algebras of order $n > 1$ are solenoidal.

An immediate generalization of the proofs in n. 5 allows to state that *all regular algebras* (total matric algebras) [Editors' note: this is written in English in the original text; see also Definition 2.3 and the comment after that.] *and all the real or complex Clifford algebras are solenoidal but not bisolenoidal.*

---

[26] See Albert cited in I [(4)], Ch. IX, n. 11.
[27] See Albert cited in [(26)], Chapt. IX, n.11.

From the two propositions it follows that *simple real or complex algebras*, i.e. direct product of a division algebra with an algebra which is regular, *of order $n > 1$ are solenoidal.*

In force of the first theorem in n. **5**, the real semi-simple algebras, i.e. direct sums of simple algebras, are solenoidal if and only if all their components are so; since there are no simple solenoidal algebras of order $1, 3, 5, 7,$[28] we conclude that *real semi-simple algebras of order $1, 3, 5, 7$ are not solenoidal.*

An algebra $\mathscr{A} = (u_1, \ldots, u_{n-m}; u_{n-m+1}, \ldots, u_n) = (u'; u'')$ of order n not semisimple has a nontrivial subalgebra $\mathscr{R} = (u'')$ of order m which is its maximal nilpotent ideal, called radical of the algebra;[29] in the case of $\mathscr{A}$, (2.14) rewrites as:

$$(u' \; u'') \begin{pmatrix} A_1 & A_2 \\ A_3 & A_4 \end{pmatrix} \begin{pmatrix} u'_{-1} \\ u''_{-1} \end{pmatrix} = u' A_1 u'_{-1} + (u' \; u'') \begin{pmatrix} 0 & A_2 \\ A_3 & A_4 \end{pmatrix} \begin{pmatrix} u'_{-1} \\ u''_{-1} \end{pmatrix} = 0,$$

and, since $\mathscr{R}$ and $\mathscr{A} - \mathscr{R}$ are disjoint, this condition imposes the vanishing of the two factors in the right hand side. It follows that: *in order that the real algebra $\mathscr{A}$ is solenoidal, $\mathscr{A} - \mathscr{R}$ has to be solenoidal.*

Since the real algebra $\mathscr{A} - \mathscr{R}$ is semisimple,[30] from the last two statements it follows that *the real algebras of orders $n - 1$, $n - 3$, $n - 5$, $n - 7$ are not solenoidal.*[31]

Algebras with cyclic radical, namely with radical of order m and index $m + 1$, are direct sum of two algebras one of which is either the algebra of $(m+1)$-potential numbers or the algebra of ternions;[32] since, by virtue of the last statement of n. 5, they are not solenoidal, *real algebras with cyclic radical are not solenoidal.* In particular, *algebras with radical of order 1 are not solenoidal.*

Bearing in mind that if $\mathscr{A} - \mathscr{R}$ is simple, its order must divide both the order of $\mathscr{A}$ and the one of $\mathscr{R}$,[33] we show that *there are no solenoidal real algebras of order 5 and 7.*

We know already that there are no semisimple, solenoidal, real algebras of orders 5, 7; thus, recalling the next to the last theorem, algebras whose radicals have orders 4, 2, 1, 0 or 6, 4, 2, 1, 0, respectively, are not solenoidal.

If there existed a real solenoidal algebra $\mathscr{A}$ of order 5 with radical $\mathscr{R}$ of order 3, the algebra $\mathscr{A} - \mathscr{R}$ of order 2 would be solenoidal and thus it would be the algebra

---

[28] There is a simple solenoidal algebra of order 9, the regular algebra of 3×3 matrices.

[29] See Albert cited in I [(4)], Ch. II, n.5.

[30] See G. Scorza, *Sopra un teorema fondamentale della teoria delle algebre*, Rend. Acc. Lincei (6), vol. 20, p. 65–72 (1934).

[31] Since semisimple, commutative algebras of odd order cannot be solenoidal, we will have that *real commutative algebras of even order with radical of order $n - 1$, $n - 3$, ..., are not solenoidal.* An analogous result holds for n odd.

[32] See G. Scorza, *Le algebre per ognuna delle quali la sottoalgebra eccezionale è potenziale*, Acc. Sc. Torino, vol. 70 (1934-35), n. 11. Let us recall that here we consider only algebras with unit.

[33] See Scorza cited in [(30)], n. 5.

of complex numbers, which is simple; in this case, its order must divide the order of $\mathscr{A}$ and this is absurd.

The same reasoning shows that the radical of a real solenoidal algebra of order 7 cannot be of order 5 and that if $\mathscr{R}$ is of order 3 then $\mathscr{A} - \mathscr{R}$ cannot be simple. Thus let us suppose that there exists a real, solenoidal algebra of order 7 with radical of order 3—which will not be cyclic—and let $\mathscr{A} - \mathscr{R}$ be direct sum of the algebra of complex numbers with itself; the multiplication table of $\mathscr{A}$ would be $\begin{pmatrix} T_1 & T_2 \\ T_3 & T_4 \end{pmatrix}$, where T_1 is the multiplication table of $\mathscr{A} - \mathscr{R}$ whose units combine according to the rules $u_1^2 = u_1$, $u_1 u_2 = u_2$, $u_2^2 = u_1$, $u_3^2 = u_3 u_3 u_4 = u_4 u_4^2 = -u_3$, [Editor's note: $u_3^2 = u_3$, $u_3 u_4 = u_4$, $u_4^2 = -u_3$], T_2 and T_3 are matrices with elements in $\mathscr{R}$ and T_4 is the multiplication table of a nilpotent algebra of order 3 whose units combine according to the rules $u_5^2 = u_6 u_5 = u_7$, $u_6^2 = \alpha u_7$ or $u_5^2 = u_7$, $u_6^2 = \alpha u_7$ (case 1), $u_5 u_6 = -u_6 u_5 = u_7$ (case 2) or it is the table of a zero algebra (case 3).[34] Then let $u = (u_1 + u_3 + \alpha u_5 + b u_6 + c u_7)$ be the unit of $\mathscr{A}$ and let us set

$$(u_2 + u_4)u_{4+i} = \sum_k m_{ik} u_{4+k}$$

$(i, k = 1, 2, 3)$ with m_{ik} real numbers.

In the first two cases we will have, respectively,

$$(u_2 + u_4)u_7 = (u_2 + u_4)u_5^2 = \left(\sum_i m_{1i} u_{4+i}\right)u_5 = \gamma u_7$$

$$(u_2 + u_4)u_7 = (u_2 + u_4)u_5 u_6 = \left(\sum_i m_{1i} u_{4+i}\right)u_6 = \gamma u_7$$

with γ real (it can also be zero). On the other hand

$$u_7 = u u_7 = (u_1 + u_3)u_7 = -(u_2 + u_4)^2 u_7 = -\gamma^2 u_7;$$

thus the real algebra at hand cannot have a radical as in case 1 or 2. Thus let us consider case 3; then

$$u_{4+i} = u u_{4+i} = (u_1 + u_3)u_{4+i} = -(u_2 + u_4)^2 u_{4+i} = -\sum_k m_{ik}\left(\sum_i m_{ik} u_{i+4}\right),$$

thus $\|m_{ik}\| + I_3 = 0$. But a real matrix of odd order cannot satisfy such an equation, thus the radical of $\mathscr{A}$ cannot be a zero-algebra of order 3; this completes the proof of the theorem.

---

[34] See Scorza cited in $^{(2)}$, § 2. [Editors's Note: it should be Scorza cited in $^{(3)}$, § 2].

2.3   Monogenicity and Total Derivability in Real and Complex Algebras, III

Article III by Michele Sce, continuation of the Notes I, II published in these "Rendiconti" pp. 30–35 and pp. 188–193 presented during the meeting of 13 March 1954 by B. Segre, member of the Academy.

7. Let x and y be elements of the algebra $\mathscr{A}$ as in n. **1** and let $y^{(k)}$ denote the partial derivative of $y(x)$ with respect to x_k; in order to have that the Pfaffian form $y\,dx$ is closed it is necessary that

$$y^{(k)}u_h - u_k y^{(h)} = 0, \qquad (h, k = 1, \ldots, n) \tag{2.17}$$

so that $y(x)$ is right totally derivable.[35] If—possibly making a basis change—we assume that u_1 is the unit of $\mathscr{A}$, from (2.17) we obtain

$$y^{(1)}(u_k u_h - u_h u_k) = 0, \qquad (h, k = 1, \ldots, n); \tag{2.18}$$

so that, if $\mathscr{A}$ is not commutative, $y^{(1)}$ is a zero-divisor and the matrix which corresponds to it in the algebra $\mathscr{A}'$, first regular representation of $\mathscr{A}$, is singular. Since this matrix is, by virtue of the total derivability, the jacobian matrix of $y(x)$ we have that *functions $y(x)$ with non-zero jacobian such that $y\,dx$ is closed are the functions totally derivable in a commutative algebra.*

Since, by virtue of the results in n. **2**, the form $y\,dx$ is co-closed if and only if the function $y(x)$ is monogenic[36] we can state that *in the commutative, solenoidal algebras, closed forms are co-closed, thus they are harmonic both in the Hodge and in the de Rham sense.*[37]

8. Let S_n be the vector space associated with $\mathscr{A}$ and let y be right monogenic in a domain D, y' a left monogenic function in a domain D'; then if V_{n-1} is a $(n-1)$-dimensional cycle contained in $D \cap D'$ and homologous to zero there, and dx^* is the adjoint of the form $dx = \sum_i u_i\,dx_i$, $(i = 1, \ldots, n)$, one has

$$\int_{V_{n-1}} y\,dx^*\,y' = 0.^{38} \tag{2.19}$$

---

[35] See A. Kriszten, *Hypercomplexe und pseudo-analytische Funktionen*, Comm. Math. Helv. v. 26 (1952), pp. 6–35, § 2; the result appears in integral form in Rizza ciated in [2], n. 11. If one denotes by dy/dx the jacobian matrix of $y(x)$ and by T the table of multiplication of $\mathscr{A}$, one can see that (2.17) may be written in the compact form $\left(\dfrac{dy}{dx}\right)_{-1} T = T_{-1}\dfrac{dy}{dx}$ which is evidently invariant with respect to changes of basis.

[36] See Kriszten cited in [35] §3; the result, though in integral form, can be found already in Rizza cited in I [2], n.7 and in Segre cited in I [2], p. 446.

[37] The statement improves the one given by Kriszten cited in [35] §4, since this Author does not take into account that monogenicity depends on the basis.

[38] See Fueter cited in I [2] and Rizza cited in [36].

Now let us suppose that for the monogenic functions of $\mathscr{A}$ there exists an integral formula of Cauchy-type. More precisely, let us assume that in $\mathscr{A}$ there exists a function $f(x, \xi)$ which, for ξ fixed, is right monogenic in x in S_n except a set $\mathscr{I}$, at most $(n-1)$-dimensional, of points in which it is not defined, so that for every $g(x)$ left monogenic in a domain D' one has

$$\int_{V_{n-1}} f(x, \xi) \, dx^* \, g(x) = g(\xi) \tag{2.20}$$

where V_{n-1} is an $(n-1)$-dimensional cycle encircling ξ and homologous to zero in D'; by virtue of the theorem just stated, V_{n-1} cannot be homologous to zero in the domain where $f(x, \xi)$ is monogenic and must contain the points in $\mathscr{I}$. In particular, (2.20) must hold when $g(x)$ is the unit of the algebra and V_{n-1} is a sphere centered at ξ and with radius r; setting

$$x_1 = \xi_1 + r \cos \varphi_1 \cdots \cos \varphi_{n-1}, \; x_2 = \xi_2 + r \sin \varphi_1 \cos \varphi_2 \cdots \cos \varphi_{n-1}, \; \ldots$$

$$\ldots, \; x_n = \xi_n + r \sin \varphi_{n-1}, \qquad (0 \le \varphi_1 < 2\pi; \; -\frac{\pi}{2} \le \varphi_i \le \frac{\pi}{2}, \ldots, i = 2, \ldots, n-1),$$

since it turns out that

$$dx^* = r^{n-2}(x - \xi) \, d\sigma,$$

where $d\sigma = \cos^{n-2} \varphi_{n-1} \cdots \cos \varphi_2 \, d\varphi_1 \cdots d\varphi_{n-1}$ is the area element of the unit sphere,[39] (2.20) gives

$$\int_S f(x, \xi) r^{n-2}(x - \xi) \, d\sigma = 1. \tag{2.21}$$

Thus *in the algebras where the function $r^{2-n}(x - \xi)^{-1}$ is right monogenic, where defined,* namely it satisfies

$$\sum_k \frac{\partial}{\partial x_k}[r^{2-n}(x - \xi)^{-1}]u_k =$$

$$= r^{-n}(x - \xi)^{-1}[(n-2)(x - \xi) + r^2 \sum_k u_k(x - \xi)^{-1}u_k] = 0, \tag{2.22}$$

we can presume that formula (2.20) holds where we have set

$$f(x, \xi) = k^{-1}r^{2-n}(x - \xi)^{-1}, \qquad k = \int_S d\sigma. \tag{2.23}$$

---

[39] See W. Nef, *Ueber eine Verallegemeinnerung des Satzes von Fatou für Potentialfunktionen,* Comme. Math. Helv., vol. 13 (1943–44), pp. 215–241, in between p. 231 and p. 232.

Since in Clifford algebras, for any nonzero divisor x one has $\sum_k u_k x u_k = -(n - 2)\bar{x}$, with $x\bar{x} = r^2$, (2.22) is satisfied and it remains to establish if (2.20), where we have set (2.23), effectively gives an integral formula in Clifford algebras.[40]

9. To obtain real solutions to the equation

$$\Omega h(x_1, \ldots, x_p) = \sum_{k=1}^{p} \alpha_k(x_1, \ldots, x_p) \frac{\partial^2 h}{\partial x_k^2} = 0 \qquad (2.24)$$

we can consider the equation

$$\Omega f(x) = 0 \qquad (2.25)$$

where f is a totally derivable function of the element $x = x_1 u_1 + \cdots + x_p u_p$ of the algebra $\mathscr{A}$ of order n ($p \leq n$). Equation (2.25) is solvable is and only if, with respect to some basis of $\mathscr{A}$, it holds

$$\sum_{k=1}^{p} \alpha_k u_k^2 = 0.^{41}$$

Thus, *if $p = n$ and α_k are constant, (2.25) is solvable in solenoidal algebras in the complex field; in particular, if (2.24) is elliptic, (2.25) is solvable in solenoidal algebras in the field of real numbers.*

The aforementioned method easily extends to partial differential equations of order greater that two; for example, to solve the equation

$$\Delta^n h(x_1, x_2) = 0$$

we can bring back to totally derivable functions in an algebra such that

$$(u_1^2 + u_2^2)^n = 0.$$

Among this type of algebras, are particularly relevant the cyclic algebras of order $2n$ whose basis $1, j, j^2, \ldots, j^{2n-1}$ satisfies the relation $(1 + j^2)^n = 0$. By setting $\omega = 1 + j^2$, we can express j through the imaginary unit and powers of ω; thus one sees that such algebras are direct product of the algebra of complex

---

[40]The integral formula given by R. Fueter, *Die Funktionentheorie der Differentialgleichungen $\Delta u = 0$ und $\Delta\Delta u = 0$ mit vier reellen Variablen*, Comment. Math. Helv., v. 7, pp. 307–330, (1934-35), n.4 for the algebra of quaternions and those given in Clifford algebras for linear systems in Clifford algebras are of the indicated form. G. B. Rizza has promised a work on general integral formulas in Clifford algebras.

[41]See P. W. Ketchum, *Analytic functions of hypercomplex variables*, Trans. Am. Math. Soc., v. 30 (1928), pp. 641–667, n. 25.

numbers and algebras of n-potential numbers with basis $1, \omega, \ldots, \omega^{n-1}$. [42] This fact ensures that these algebras are solenoidal and such that totally derivable functions $y(x_0, \ldots, x_{2^n-1})$ are harmonic; moreover, a simple inspection of the jacobian matrix which, by definition of total derivability is of the form

$$\begin{pmatrix} \alpha_0 & 0 & 0 & \cdots \\ \alpha_1 & \alpha_0 & 0 & \cdots \\ \alpha_2 & \alpha_1 & \alpha_0 & \cdots \\ \vdots & \vdots & \vdots & \cdots \end{pmatrix} \quad \text{with } \alpha_k = \begin{pmatrix} a_{2k} & -a_{2k+1} \\ a_{2k+1} & a_{2k} \end{pmatrix} \quad (2.26)$$

ensures that every y is a harmonic function of all pairs x_{2k}, x_{2k+1}. It is worthwhile to note that monogenic functions which are not totally derivable are not even harmonic.

10. Let us now consider noncommutative algebras $\mathscr{A}_n$ of order $2n$ whose basis

$$1, i, \omega, i\omega, \ldots, \omega^{n-1}, i\omega^{n-1}$$

satisfies the relations

$$i^2 = -1, \quad \omega^n = 0, \quad \omega i + i\omega = 0; \text{ [43]}$$

we can write any element a in $\mathscr{A}_n$ in the form

$$a = \alpha_0 + \alpha_1\omega + \cdots + \alpha_{n-1}\omega^{n-1} \quad (2.27)$$

where $\alpha_k = a_{2k} + ia_{2k+1}$ behave among them like ordinary complex numbers, while

$$\alpha_k\omega = \omega\bar{\alpha}_k. \quad (2.28)$$

Besides the algebras $\mathscr{A}_n$, we will consider the algebra $\mathscr{Q}_n$ of order $4n$ which can be obtained by maintaining condition $\omega^n = 0$ and assuming that α_k in (2.27) and (2.28) behave among them like ordinary quaternions.

---

[42] For $n = 2$ one has the algebra studied by Sobrero cited in I [(1)], which, from another point of view Ketchum already crossed in Ketchum cited in [(41)], n. 31. [Editors' Note: the algebra studied by Sobrero and Ketchum is commutative and corresponds to the case $i\omega = \omega i$, i.e. to algebra LXXIX in Scorza's classification [32]. The algebra that Sce is studying here is noncommutative and corresponds to algebra LXXXI in [32].] In this case, the transformations from one basis to another can be expressed by the relations $2i = 3j + j^3$, $\omega = a(1 + j^2) + b(j + j^3)$, $i\omega = -b(1 + j^2) + a(1 + j^3)$ with a, b arbitrary real numbers (non both zero); to the best of our knowledge, these general relations have never been considered.

[43] For $n = 2$ one has the algebra LXXXI of n. **5**.

Taking into account the expression of the product of two elements $a = \sum_k \alpha_k \omega^k$, $b = \sum_k \beta_k \omega^k$

$$ab = \sum_k \alpha_r \tilde{\beta}_s \omega^k, \quad (r + s = k),$$

$\tilde{\beta}_s = \bar{\beta}_s$ or β_s according to the fact that $r = k - s$ is even or odd, with long but not difficult calculations one can establish the following results:

the elements in the center of an algebra $\mathscr{A}_n$ or $\mathscr{Q}_n$ defined via (2.27) and (2.28) and $\omega^n = 0$ have the form $\sum \alpha_{2k} \omega^{2k}$ with $\alpha_{2k} = \bar{\alpha}_{2k}$.

The zero divisors are of the form $\omega^i a$ thus they are all and the only nilpotent elements in the algebra; these latter constitute the radical, which is of index n and order $2(n - 1)$ for $\mathscr{A}_n$ and $4(n - 1)$ for $\mathscr{Q}_n$.

The only idempotent not nilpotent is the unit thus $\mathscr{A}_n$ and $\mathscr{Q}_n$ are algebras completely primary.

If one considers only the case $n = 2$, as we shall do, from the first statement one obtains immediately that $\mathscr{A}_2$ and $\mathscr{Q}_2$ are normal. Moreover, setting $\bar{a} = \bar{\alpha}_0 - \alpha_1 \omega$, if we say that the *norm* of a is the real number $a\bar{a} = \alpha_0 \bar{\alpha}_0$, we see that *the zero divisors are all the elements with zero norm and only them.*

By setting $y = \alpha + \beta \omega$ and $x = \xi + \eta \omega$, *monogenic functions $y(x)$ are also harmonic in the components of ξ.*

In fact, it is known[44] that the monogenicity condition for $y(x)$ may be written as $Dy = 0$ where D is an operator that behaves like an element in the algebra; since the norm of D is the laplacian associated with the components of ξ, by applying to $Dy = 0$ the operator $\bar{D}$ we obtain the result.

11. The considerations that we made so far, even though maybe not uninteresting, would be in need of being deepened if one wishes to deduce more concrete results; however, it is our belief that such results can be obtained only in special type of algebras like, for example, $\mathscr{A}_n$.

About these latter algebras we point out that their zero divisors, in the representative space S_{2n}, form a linear space $S_{2(n-1)}$; thus, in this case, the study of the variety of the zero divisors—necessary preliminary to look for integral formulas—is trivial. The difficulty in this type of problems is the lack of concrete examples of monogenic functions, especially in the case of noncommutative algebras;[45] for example, in $\mathscr{A}_2$ neither the powers nor the exponential are monogenic function, and we can only say that such functions are of the form

$$w = u(x_1, x_2) + \frac{v(x_1, x_2)}{x_2}(x_2 i + x_3 \omega + x_4 i \omega)$$

---

[44] See Segre cited in cited in I $^{(2)}$ p. 442.

[45] It is less difficult in commutative solenoidal algebras, since in these algebras the most common functions in analysis are totally derivable.

with $u + iv$ holomorphic. In the algebra of quaternions, from a similar property one deduces that Δw (and in particular Δx^n) are monogenic functions; since such a result is not valid in $\mathscr{A}_2$, there is the problem of knowing if in $\mathscr{A}_2$ there is a differential operator Ω such that Ωw are monogenic.

Since all these problems appear to be connected among them, an answer, even partial, could shed light on the whole question: this is what we hope to do in another work.

2.4 Comments and Historical Remarks

In this section we are revising, also providing examples, the concepts contained in the previous sections i.e. in the original papers [27–29]. We also add some historical remarks which seem to be nowadays forgotten.

The interest in theories of functions in algebras other than the algebra of complex numbers and generalizing the complex holomorphic functions started after the study of these algebras in the classical works of Gauss, Hamilton, Hankel, Frobenius and goes back to the end of the nineteenth century (see [40] for a list of references). It then continued with the work of Lanczos [17], who considered the generalization to quaternions back in 1919, but his work was mostly unknown until [18], and with the PhD dissertation of Ketchum, see [14]. It was in the early thirties when the study of quaternionic functions started systematically with the works of Moisil [23] and Fueter and in the forties, Krylov [16] and Meilikhson [22] studied the notion of quaternionic differentiability.

It is interesting that in the thirties and forties, various authors considered functions with values in algebras, for example Ward [40] who developed his PhD dissertation on the theory of analytic functions in associative algebras, Nef [24], Spampinato [38], Sobrero [37] and also Fueter [7–9] and Haefeli [11]. The interest in these studies continued even later as shown by the works of Kriszten [15] and Rizza [26].

Michele Sce's works considered in this chapter insert in this field of researches. He knew rather well the existing literature and despite the fact that the circulation of the journals, and consequently papers, was more limited in the fifties, his knowledge of the available works was complete. Sce notes in his papers, that some function theories were already developed, specifically the theory of hyperholomorphic (or hyperdifferentiable or monogenic) functions over quaternions, bicomplex numbers, Clifford algebras. In the quaternionic case, the analog of holomorphic functions are the Cauchy–Fueter regular functions, so-called since it was Fueter and his school who developed this function theory.

In his work Sce, as well as a few other authors, quotes the work of Moisil [23] published in 1931, mentioning that this author was using the term monogenic, instead of regular, functions. But the history of the birth of regularity on quaternions is more complicated. For example (as we mentioned before and without any claim to historical completeness), Lanczos already developed this approach in his 1919 dissertation [17], Fueter presented some of his results at the International Congress of Mathematicians in 1928, [6], and Iwanenko and Nikolsky already considered the case of biquaternions in 1930, [13]. It is also likely that, at that time, other Russian researchers were working in this framework. Thus, we believe that it is fair to say that various authors, more or less at the same time, were considering quaternionic functions and a notion of holomorphicity in this context.

The second four dimensional case (over $\mathbb{R}$), namely the one of bicomplex numbers, was started by Scorza Dragoni in [35] and, after the monograph [25], it has attracted attention in more recent times, see [20]. We also note that another four dimensional case, the one of bidual numbers studied by Sobrero [37] has been basically abandoned. We note that we kept the name "bidual" to be consistent with the terminology adopted by the Italian school, even though this algebra is nothing but the complex Grassmann with one generator $\mathfrak{f}_1$ such that $\mathfrak{f}_1^2 = 0$. Algebras of order up to four have been studied in [30–34].

The case of functions Clifford algebra valued is widely studied in the literature, starting with the celebrated monograph [3] which has been followed by several other books and hundreds of papers. However, one should notice that the notion of monogenicity treated in this chapter is given for functions from (a subset of) an algebra to itself. This is not the case treated in [3] and subsequent literature, where the functions have values in a Clifford algebra but are defined on Euclidean space identified with the set of paravectors or of vectors in the algebra. It is remarkable that Sce already considered this class of functions, as we shall see in Chap. 5, in relation with the celebrated Fueter theorem nowadays known as Fueter-Sce-Qian theorem. These functions were eventually considered by Iftimie [12] thus denoting that also the Romanian school was continuing the studies in hypercomplex analysis.

Below, we will use some examples to illustrate what is presented in the previous sections and to compare the various concepts. In our examples, we will discuss some particular choices of algebras of order up to four.

We will use standard terminology and notation, like A^T (instead of A_{-1} used by Sce) to denote the transpose of a matrix A and, in particular, of a vector. By $\mathscr{A}$ we denote a real or complex algebra of order n and by $\mathscr{A}'$, $\mathscr{A}''$ the algebras of $n \times n$ matrices which are the first and second regular representation of $\mathscr{A}$.

Even though the basic notions and terminology about algebras are well known, we repeat some preliminaries for the sake of completeness.

Definition 2.1 An algebra over a field F is a set $\mathscr{A}$ such that $\mathscr{A}$ is a vector space over F and there is a F-bilinear mapping from $\mathscr{A} \times \mathscr{A} \mapsto \mathscr{A}$, $(a, b) \mapsto ab$, i.e.

$$k(ab) = (ka)b = a(kb), \qquad \text{for all } k \in F, \ a, b \in \mathscr{A}.$$

This bilinear map is called multiplication.

In particular, an algebra is called associative (resp. commutative) if the multiplication is associative (resp. commutative).

Even though it is not specified in Sce's papers, all the algebras considered are associative. This was a standard assumption in the papers written in Italy at that time, unless otherwise specified, and it appears also from the calculations performed in the manuscripts.

Definition 2.2 We say that $\mathscr{A}$ has order n if there exist $u_1, \ldots, u_n \in \mathscr{A}$ such that every $a \in \mathscr{A}$ can be expressed in a unique way as

$$a = a_1 u_1 + \cdots + a_n u_n, \qquad a_i \in F, \ i = 1, \ldots, n.$$

We recall that, in this Chapter, we consider associative algebras and that an algebra $\mathscr{A}$ is called division algebra if it is, as a ring, a division ring.
We now state more definitions that are used in the book.

Definition 2.3 An algebra is called regular if it is isomorphic to an algebra of $m \times m$ matrices.

In **6** Sce uses the term *regular* also referring to the English terminology *total matric algebra* which, however, seems to be not anymore in use.

Definition 2.4 An element $a \in \mathscr{A}$ is called nilpotent if $a^r = 0$ for some $r \in \mathbb{N}$ and the least such r is called index of a. Moreover, a is called properly nilpotent if both ya and ay are zero o nilpotent for every $y \in \mathscr{A}$.

Definition 2.5 The set $\mathscr{R}$ consisting of zero and of all properly nilpotent elements is called radical of $\mathscr{A}$.
An algebra is called semi-simple if its radical is the zero ideal.
An algebra is called simple if its only proper ideal is the zero ideal and $\mathscr{A}$ is not a zero algebra of order 1, namely $\mathscr{A}$ is not such that $ab = 0$ for every $a, b \in \mathscr{A}$.

It is also useful to recall that

Theorem 2.1 *Let $\mathscr{N}$ be an ideal of an algebra $\mathscr{A}$. Then $\mathscr{A} \setminus \mathscr{N}$ is semi-simple if and only if $\mathscr{N}$ is the radical ideal of $\mathscr{A}$.*

As explained in Sect. 2.1, the notion of total differentiability has been introduced by Spampinato in [38]. The development of a function theory starting from this definition was not really developed. In the case of left (resp. right) derivability, meant as the existence of the limit of the left (resp. right) difference quotient

$$(q - q_0)^{-1}(f(q) - f(q_0)) \qquad (f(q) - f(q_0))(q - q_0)^{-1}$$

where q, q_0 are quaternions, one obtains affine functions only. This fact was proved by Meilikhson in [22], but the interested reader may find a proof in Sudbery's paper [39] which made the result commonly known.

It was realized only at a later stage, see for examples the works [10, 19, 21] by Gürlebeck, Malonek, Shapiro and others, that in order to obtain a meaningful class of functions one needs to construct differently the quotients. By taking these suitable difference quotients, the notion of differentiability that one obtains coincides with the notion of monogenicity, in analogy with the complex case.

In order to write the notion of total derivability used in this work, we recall the next definition (see (2.1), (2.2) and [1, 2]):

Definition 2.6 Let $\mathscr{A}$ be a real or complex algebra of order n, with unit and let $u = (u_1, \ldots, u_n)$ be a vector containing the ordered elements in a given basis of $\mathscr{A}$. We say that the algebras $\mathscr{A}'$, $\mathscr{A}''$ are the *first and second regular representation* of $\mathscr{A}$ if their elements the are order n matrices X', X'' defined, for any $x \in \mathscr{A}$, by the relations

$$xu = uX'$$

$$ux = u(X'')^T.$$

Definition 2.7 Let $\mathscr{A}$ be an algebra with basis $u_1, \ldots, u_n$, $x \in \mathscr{A}$ and $y : \mathscr{A} \longrightarrow \mathscr{A}$. Let $u = (u_1, \ldots, u_n)$ and let $\xi = (\xi_1, \ldots, \xi_n)$, $\eta = (\eta_1, \ldots, \eta_n)$ be the coordinates of x, y, respectively, with respect to the given basis, i.e.

$$x = \xi u^T,$$

$$y = \eta u^T.$$

If y is derivable, that is, all the components of η are derivable with respect to the components of ξ, we say that y is right (resp. left) totally derivable if the jacobian $d\eta/d\xi$ belongs to $\mathscr{A}'$ (resp. the transpose of the jacobian belongs to $\mathscr{A}''$).

Remark 2.1 The notion of right or left total derivability is designed on the notion of right or left differentiability, in the standard sense. In fact, let us consider a function y with values in an algebra with unit $\mathscr{A}$, where a basis $u_1, \ldots, u_n$ is fixed:

$$y(x) = y_1(x)u_1 + \cdots + y_n(x)u_n,$$

where $x = x_1u_1 + \cdots + x_nu_n$. Note that x_i, y_i, $i = 1, \ldots, n$ are real or complex and x varies in an open set of $\mathscr{A}$ of when we identify $\mathscr{A}$ with $\mathbb{R}^n$ (or $\mathbb{C}^n$). If we assume that the functions y_ℓ admit derivatives with respect to x_i and we set

$$dx = dx_1u_1 + \cdots + dx_nu_n, \qquad dy = dy_1u_1 + \cdots + dy_nu_n,$$

with

$$dy_\ell = \frac{\partial y_\ell}{\partial x_1}dx_1 + \cdots + \frac{\partial y_\ell}{\partial x_n}dx_n, \qquad \ell = 1, \ldots, n,$$

then the function y is left differentiable or totally derivable on the left if there exists a function $z(x)$ such that

$$dy = dx\, z(x),$$

y is right differentiable or totally derivable on the right if there exists a function $z(x)$ such that

$$dy = z(x)\, dx, \tag{2.29}$$

for every dx. Writing $z(x) = z_1(x)u_1 + \cdots + z_n(x)u_n$ and setting

$$u_i u_j = \sum_{\ell=1}^{n} \gamma_{ij\ell} u_\ell,$$

we have that (2.29) becomes

$$\sum_{\ell=1}^{n} dy_\ell u_\ell = \sum_{i,j,\ell=1}^{n} \gamma_{ij\ell} z_i dx_j u_\ell.$$

By equating the coefficients in front of the units u_ℓ, we deduce:

$$dy_\ell = \sum_{i,j=1}^{n} \gamma_{ij\ell} z_i dx_j, \qquad \ell = 1, \dots, n.$$

Since $dy_\ell = \sum_{j=1}^{n} \dfrac{\partial y_\ell}{\partial x_j} dx_j$ and the differentials dx_i are independent we obtain

$$\sum_{i=1}^{n} \gamma_{ij\ell} z_i = \frac{\partial y_\ell}{\partial x_j}, \qquad j, \ell = 1, \dots, n. \tag{2.30}$$

Thus the total derivability on the right (2.29) is equivalent to (2.30). The left hand side of (2.30) gives the entries $x'_{\ell j}$ of the matrix X' of the first regular representation of z. Thus (2.30) expresses the fact that the Jacobian matrix $(\partial y_\ell / \partial x_j)$ belongs to $\mathscr{A}'$.

In Definition 2.7, x is varying in the whole algebra, but when a topology can be defined (for example identifying the elements in a real algebra of order n with vectors in $\mathbb{R}^n$) we can consider an open set U in $\mathscr{A}$ and have the notion of right or left totally derivable function on U with values in $\mathscr{A}$. The notion implies that the jacobian matrix satisfies suitable symmetries, as shown in the following examples.

Example 2.1 Let us consider the case of the real algebra of complex numbers. Then a basis is given for example by $\{1, i\}$ with $i^2 = -1$, so that $u = (1 \ \ i)$. Then $x = x_1 + ix_2$ and

$$xu = (x_1 + ix_2 \ \ ix_1 - x_2) = (1 \ \ i) \begin{pmatrix} x_1 & -x_2 \\ x_2 & x_1 \end{pmatrix}.$$

In this commutative case the two representations $\mathscr{A}'$ and $\mathscr{A}''$ coincide. The jacobian matrix

$$\begin{pmatrix} \dfrac{\partial y_1}{\partial x_1} & \dfrac{\partial y_1}{\partial x_2} \\ \dfrac{\partial y_2}{\partial x_1} & \dfrac{\partial y_2}{\partial x_2} \end{pmatrix}$$

belongs to $\mathscr{A}'$ if and only if

$$\frac{\partial y_1}{\partial x_1} = \frac{\partial y_2}{\partial x_2}, \qquad \frac{\partial y_1}{\partial x_2} = -\frac{\partial y_2}{\partial x_1}$$

namely if and only if the Cauchy–Riemann conditions are satisfied.

We know from the general theory that total derivability is independent of the choice of a basis. In this specific example, let us choose the basis $\{1, -i\}$. Then $x = x_1 - ix_2$

$$xu = (x_1 - ix_2 \ \ -ix_1 - x_2) = (1 \ \ -i)\begin{pmatrix} x_1 & -x_2 \\ x_2 & x_1 \end{pmatrix}.$$

Thus the conclusion is as above.

Example 2.2 Let us consider the algebra of dual numbers, namely the algebra generated by u_1, u_2 satisfying $u_1^2 = u_1$, $u_2^2 = 0$, $u_1 u_2 = u_2 u_1 = u_2$. Let $x = x_1 u_1 + x_2 u_2$, $u = (u_1 \ u_2)$. Then, since the algebra is commutative the right and left representations coincide and follow from

$$xu = (x_1 u_1 + x_2 u_2 \ \ x_1 u_2) = (u_1 \ u_2)\begin{pmatrix} x_1 & 0 \\ x_2 & x_1 \end{pmatrix}.$$

Thus the condition of right and left total derivability of $y(x) = y_1(x)u_1 + y_2(x)u_2$ is then

$$\frac{\partial y_1}{\partial x_1} = \frac{\partial y_2}{\partial x_2}, \qquad \frac{\partial y_1}{\partial x_2} = 0.$$

To conclude the examples in the case of algebras of second order, we consider the case of hyperbolic numbers:

Example 2.3 Let us consider the algebra of hyperbolic numbers, namely the algebra generated by u_1, u_2 satisfying $u_1^2 = u_1$, $u_2^2 = u_1$, $u_1 u_2 = u_2 u_1 = u_2$. Let

$x = x_1u_1 + x_2u_2$, $u = (u_1 \; u_2)$. Due to the commutative setting, the left and right representations follow from

$$xu = ux = (x_1u_1 + x_2u_2 \; x_1u_2 + x_2u_1) = (u_1 \; u_2) \begin{pmatrix} x_1 & x_2 \\ x_2 & x_1 \end{pmatrix}.$$

Thus the condition of total derivability, both left and right, is expressed by

$$\frac{\partial y_1}{\partial x_1} = \frac{\partial y_2}{\partial x_2}, \quad \frac{\partial y_1}{\partial x_2} = \frac{\partial y_2}{\partial x_1}.$$

Example 2.4 We now consider the case of an algebra of the third order, specifically the algebra of ternions which is defined as the real algebra of upper triangular 2×2 matrices. As a basis of the algebra, we choose

$$u_1 = \begin{pmatrix} 1 & 0 \\ 0 & 0 \end{pmatrix}, \quad u_2 = \begin{pmatrix} 0 & 0 \\ 0 & 1 \end{pmatrix}, \quad u_3 = \begin{pmatrix} 0 & 1 \\ 0 & 0 \end{pmatrix}.$$

The multiplication rules are

$$u_1^2 = u_1, \quad u_2^2 = u_2, \quad u_3^2 = 0, \quad u_1u_3 = u_3, \quad u_3u_2 = u_3,$$

$$u_1u_2 = u_2u_1 = u_2u_3 = u_3u_1 = 0.$$

Setting $x = x_1u_1 + x_2u_2 + x_3u_3$, the first representation can be computed from

$$xu = (x_1u_1 \; x_2u_2 + x_3u_3 \; x_1u_3) = (u_1 \; u_2 \; u_3) \begin{pmatrix} x_1 & 0 & 0 \\ 0 & x_2 & 0 \\ 0 & x_3 & x_1 \end{pmatrix}$$

while the second representation follows from

$$ux = (x_1u_1 + x_3u_3 \; x_2u_2 \; x_2u_3) = (u_1 \; u_2 \; u_3) \begin{pmatrix} x_1 & 0 & 0 \\ 0 & x_2 & 0 \\ x_3 & 0 & x_2 \end{pmatrix}.$$

Thus the conditions of right total derivability are expressed by

$$\frac{\partial y_1}{\partial x_2} = \frac{\partial y_1}{\partial x_3} = \frac{\partial y_2}{\partial x_1} = \frac{\partial y_2}{\partial x_3} = \frac{\partial y_3}{\partial x_1} = 0, \quad \frac{\partial y_1}{\partial x_1} = \frac{\partial y_3}{\partial x_3},$$

while the left total derivability corresponds to

$$\frac{\partial y_1}{\partial x_2} = \frac{\partial y_1}{\partial x_3} = \frac{\partial y_2}{\partial x_1} = \frac{\partial y_2}{\partial x_3} = \frac{\partial y_3}{\partial x_2} = 0, \quad \frac{\partial y_2}{\partial x_2} = \frac{\partial y_3}{\partial x_3}.$$

Example 2.5 A four order algebra which has been widely studied from the point of view of a function theory on it is the one of bicomplex numbers $\mathbb{BC}$ with respect to the basis 1, i, j, k so that $x = x_1 + x_2 i + x_3 j + x_4 k$. We recall that $i^2 = j^2 = -1$, $ij = ji = k$. It is a commutative algebra, so that right and left total differentiability coincide. Since

$$xu = (x_1 + x_2 i + x_3 j + x_4 k, \ x_1 i - x_2 + x_3 k - x_4 j, \ x_1 j + x_2 k - x_3 - x_4 i, \ x_1 k - x_2 j - x_3 i + x_4)$$

$$= (1 \ i \ j \ k) \begin{pmatrix} x_1 & -x_2 & -x_3 & x_4 \\ x_2 & x_1 & -x_4 & -x_3 \\ x_3 & -x_4 & x_1 & -x_2 \\ x_4 & x_3 & x_2 & x_1 \end{pmatrix}.$$

Thus, the function $y(x)$ is left or right totally differentiable if and only if the jacobian matrix $\left(\dfrac{\partial y_i}{\partial x_k}\right)$ satisfies the conditions

$$\begin{aligned}
\frac{\partial y_1}{\partial x_1} &= \frac{\partial y_2}{\partial x_2} = \frac{\partial y_3}{\partial x_3} = \frac{\partial y_4}{\partial x_4}, \\
\frac{\partial y_1}{\partial x_2} &= -\frac{\partial y_2}{\partial x_1} = \frac{\partial y_3}{\partial x_4} = -\frac{\partial y_4}{\partial x_3}, \\
\frac{\partial y_1}{\partial x_3} &= \frac{\partial y_2}{\partial x_4} = -\frac{\partial y_3}{\partial x_1} = -\frac{\partial y_4}{\partial x_2}, \\
\frac{\partial y_1}{\partial x_4} &= -\frac{\partial y_2}{\partial x_3} = -\frac{\partial y_3}{\partial x_2} = \frac{\partial y_4}{\partial x_1}.
\end{aligned} \tag{2.31}$$

Example 2.6 Let us consider the noncommutative case of quaternions $\mathbb{H}$ with respect to the basis 1, i, j, k so that $x = x_1 + x_2 i + x_3 j + x_4 k$. We recall that $i^2 = j^2 = -1$, $ij = -ji = k$. We then have

$$xu = (x_1 + x_2 i + x_3 j + x_4 k, \ x_1 i - x_2 - x_3 k + x_4 j, \ x_1 j + x_2 k - x_3 - x_4 i, \ x_1 k - x_2 j + x_3 i - x_4)$$

$$= (1 \ i \ j \ k) \begin{pmatrix} x_1 & -x_2 & -x_3 & -x_4 \\ x_2 & x_1 & -x_4 & x_3 \\ x_3 & x_4 & x_1 & -x_2 \\ x_4 & -x_3 & x_2 & x_1 \end{pmatrix}.$$

Thus, the function $y(x)$ is right totally differentiable if and only if the jacobian matrix $\left(\dfrac{\partial y_i}{\partial x_k}\right)$ satisfies the conditions

$$\frac{\partial y_1}{\partial x_1} = \frac{\partial y_2}{\partial x_2} = \frac{\partial y_3}{\partial x_3} = \frac{\partial y_4}{\partial x_4},$$

$$\frac{\partial y_1}{\partial x_2} = -\frac{\partial y_2}{\partial x_1} = \frac{\partial y_3}{\partial x_4} = -\frac{\partial y_4}{\partial x_3},$$

$$\frac{\partial y_1}{\partial x_3} = -\frac{\partial y_3}{\partial x_1} = \frac{\partial y_4}{\partial x_2} = -\frac{\partial y_2}{\partial x_4},$$

$$\frac{\partial y_1}{\partial x_4} = \frac{\partial y_2}{\partial x_3} = -\frac{\partial y_3}{\partial x_2} = -\frac{\partial y_4}{\partial x_1}.$$

Analogously, we have

$$ux = (x_1+x_2i+x_3j+x_4k,\ x_1i-x_2+x_3k-x_4j,\ x_1j-x_2k-x_3+x_4i,\ x_1k+x_2j-x_3i-x_4)$$

$$= (1\ i\ j\ k)\begin{pmatrix} x_1 & -x_2 & -x_3 & -x_4 \\ x_2 & x_1 & x_4 & -x_3 \\ x_3 & -x_4 & x_1 & x_2 \\ x_4 & x_3 & -x_2 & x_1 \end{pmatrix}.$$

Thus the left total derivability conditions are:

$$\frac{\partial y_1}{\partial x_1} = \frac{\partial y_2}{\partial x_2} = \frac{\partial y_3}{\partial x_3} = \frac{\partial y_4}{\partial x_4},$$

$$\frac{\partial y_1}{\partial x_2} = -\frac{\partial y_2}{\partial x_1} = -\frac{\partial y_3}{\partial x_4} = \frac{\partial y_4}{\partial x_3},$$

$$\frac{\partial y_1}{\partial x_3} = -\frac{\partial y_3}{\partial x_1} = -\frac{\partial y_4}{\partial x_2} = \frac{\partial y_2}{\partial x_4},$$

$$\frac{\partial y_1}{\partial x_4} = -\frac{\partial y_2}{\partial x_3} = \frac{\partial y_3}{\partial x_2} = -\frac{\partial y_4}{\partial x_1}.$$

It is also clear that the $12 = n^2 - n$ conditions arise from imposing that the $16 = n^2$ entries depend on $4 = n$ parameters. These are definitely different from the conditions expressing the right or left monogenicity, as we shall see below.

Example 2.7 Sce made a comment on the possible interest of the algebra LXXXI in the classification given by Scorza in [34], so we consider also this case. According to n. **10**, the basis of the algebra can be written as $u = (1,\ i,\ \omega,\ i\omega)$ with $i^2 = -1$,

$\omega^2 = 0, i\omega + \omega i = 0$. Setting $x = x_1 + ix_2 + \omega x_3 + i\omega x_4$, easy computations show that

$$xu = (1, \, i, \, \omega, \, i\omega) \begin{pmatrix} x_1 & -x_2 & 0 & 0 \\ x_2 & x_1 & 0 & 0 \\ x_3 & x_4 & x_1 & -x_2 \\ x_4 & -x_3 & x_2 & x_1 \end{pmatrix},$$

while

$$ux = (1, \, i, \, \omega, \, i\omega) \begin{pmatrix} x_1 & -x_2 & 0 & 0 \\ x_2 & x_1 & 0 & 0 \\ x_3 & -x_4 & x_1 & x_2 \\ x_4 & x_3 & -x_2 & x_1 \end{pmatrix}.$$

We deduce that the function $y(x)$ is right totally differentiable if and only if the jacobian matrix $\left(\dfrac{\partial y_i}{\partial x_k} \right)$ satisfies the conditions

$$\frac{\partial y_1}{\partial x_1} = \frac{\partial y_2}{\partial x_2} = \frac{\partial y_3}{\partial x_3} = \frac{\partial y_4}{\partial x_4},$$

$$\frac{\partial y_1}{\partial x_3} = \frac{\partial y_1}{\partial x_4} = \frac{\partial y_2}{\partial x_3} = \frac{\partial y_2}{\partial x_4} = 0$$

$$\frac{\partial y_1}{\partial x_2} = -\frac{\partial y_2}{\partial x_1} = \frac{\partial y_3}{\partial x_4} = -\frac{\partial y_4}{\partial x_3},$$

$$\frac{\partial y_3}{\partial x_1} = -\frac{\partial y_4}{\partial x_2},$$

$$\frac{\partial y_3}{\partial x_2} = \frac{\partial y_4}{\partial x_1}.$$

while the left totally differentiability conditions are

$$\frac{\partial y_1}{\partial x_1} = \frac{\partial y_2}{\partial x_2} = \frac{\partial y_3}{\partial x_3} = \frac{\partial y_4}{\partial x_4},$$

$$\frac{\partial y_1}{\partial x_3} = \frac{\partial y_1}{\partial x_4} = \frac{\partial y_2}{\partial x_3} = \frac{\partial y_2}{\partial x_4} = 0$$

$$\frac{\partial y_1}{\partial x_2} = -\frac{\partial y_2}{\partial x_1} = -\frac{\partial y_3}{\partial x_4} = \frac{\partial y_4}{\partial x_3},$$

$$\frac{\partial y_3}{\partial x_1} = \frac{\partial y_4}{\partial x_2},$$

$$\frac{\partial y_3}{\partial x_2} = -\frac{\partial y_4}{\partial x_1}.$$

We now turn to the notions of right or left monogenicity that are made explicit in the examples below, computed again in the cases considered above. We recall that a function $y = y(x)$ with values in $\mathscr{A}$ is said to be right monogenic (see (2.3) and [36]) if

$$u \left(\frac{dy}{dx} \right) u^T = 0$$

or left monogenic if (see (2.3))

$$u \left(\frac{dy}{dx} \right)^T u^T = 0.$$

By explicitly writing these two conditions using the operator

$$D = \sum_{i=1}^{n} u_i \frac{\partial}{\partial x_i}$$

applied to $y(x) = \sum_{i=\ell}^{n} u_\ell y_\ell(x)$, it is clear that the second condition can be expressed as $Dy = 0$ while the second one is

$$\sum_{i=1}^{n} u_j \frac{\partial y_j}{\partial x_i} u_i = yD = 0$$

(where the notation of writing D on the right means that the units in D are written on the right).

Remark 2.2 The reader may wonder if the right and left monogenicity conditions are related by transposition of matrices. However

$$\left(u \frac{dy}{dx} u^T \right)^T = 0$$

equals

$$(u^T)^T \left(\frac{dy}{dx} \right)^T u^T = u \left(\frac{dy}{dx} \right)^T u^T$$

only if $\mathscr{A}$ is commutative. And in fact, in this case the two notions coincide.

Example 2.8 In the complex case, the notion of monogenicity (left or right, since we work in a commutative setting) is expressed by

$$(1 \; i) \begin{pmatrix} \dfrac{\partial y_1}{\partial x_1} & \dfrac{\partial y_1}{\partial x_2} \\ \dfrac{\partial y_2}{\partial x_1} & \dfrac{\partial y_2}{\partial x_2} \end{pmatrix} \begin{pmatrix} 1 \\ i \end{pmatrix} = \left(\frac{\partial y_1}{\partial x_1} + i \frac{\partial y_2}{\partial x_1} \right) + \left(\frac{\partial y_1}{\partial x_2} + i \frac{\partial y_2}{\partial x_2} \right) i = 0$$

which translates into the Cauchy–Riemann equations

$$\frac{\partial y_1}{\partial x_1} - \frac{\partial y_2}{\partial x_2} = 0, \qquad \frac{\partial y_2}{\partial x_1} + \frac{\partial y_1}{\partial x_2} = 0,$$

so the notion, as expected, corresponds to the one of holomorphicity. It is immediate to verify that one obtains exactly the same conditions if one takes the transpose of the jacobian.

Example 2.9 Another case of an algebra of second order that we considered above is the one of dual numbers. Using the basis in Example 2.2, the right (and left) monogenicity conditions are expressed by

$$\frac{\partial y_1}{\partial x_1} = 0, \qquad \frac{\partial y_1}{\partial x_2} + \frac{\partial y_2}{\partial x_1} = 0. \tag{2.32}$$

In the case of hyperbolic numbers we have:

Example 2.10 Using the basis of hyperbolic numbers given in Example 2.3, the left (and right) monogenicity conditions are expressed by

$$\frac{\partial y_1}{\partial x_1} + \frac{\partial y_2}{\partial x_2} = 0$$
$$\tag{2.33}$$
$$\frac{\partial y_2}{\partial x_1} + \frac{\partial y_1}{\partial x_2} = 0.$$

We note that in the literature one may find the above equations written with different signs. In fact, (2.33) expresses the fact that $y_1 u_1 + y_2 u_2$ is in the kernel of $u_1 \dfrac{\partial y_1}{\partial x_1} + u_2 \dfrac{\partial y_2}{\partial x_2}$, whereas in [20] one finds the conditions characterizing the kernel of $u_1 \dfrac{\partial y_1}{\partial x_1} - u_2 \dfrac{\partial y_2}{\partial x_2}$. The two function theories so obtained are different but equivalent.

Example 2.11 In the case of ternions, using the basis previously introduced, see Example 2.4, the right and left monogenicity conditions are expressed, respectively, by

$$\frac{\partial y_1}{\partial x_1} = \frac{\partial y_2}{\partial x_2} = 0, \qquad \frac{\partial y_3}{\partial x_2} + \frac{\partial y_1}{\partial x_3} = 0,$$

and

$$\frac{\partial y_1}{\partial x_1} = \frac{\partial y_2}{\partial x_2} = 0, \qquad \frac{\partial y_2}{\partial x_3} + \frac{\partial y_3}{\partial x_1} = 0.$$

Example 2.12 For the algebra $\mathbb{BC}$ the monogenicity conditions (left or right) are expressed by $u\left(\dfrac{dy}{dx}\right)u^T = 0$ and taking into account the multiplication rules we obtain

$$\frac{\partial y_1}{\partial x_1} - \frac{\partial y_2}{\partial x_2} - \frac{\partial y_3}{\partial x_3} + \frac{\partial y_4}{\partial x_4} = 0$$

$$\frac{\partial y_1}{\partial x_2} + \frac{\partial y_2}{\partial x_1} - \frac{\partial y_3}{\partial x_4} - \frac{\partial y_4}{\partial x_3} = 0$$

$$\frac{\partial y_1}{\partial x_3} - \frac{\partial y_2}{\partial x_4} + \frac{\partial y_3}{\partial x_1} - \frac{\partial y_4}{\partial x_2} = 0 \qquad (2.34)$$

$$\frac{\partial y_1}{\partial x_4} + \frac{\partial y_2}{\partial x_3} + \frac{\partial y_3}{\partial x_2} + \frac{\partial y_4}{\partial x_1} = 0.$$

If the function $y(x)$ is totally derivable, then it is also monogenic, since (2.31) imply that all the equations in (2.34) are identities, already with the basis $\{1, i, j, k\}$.

Example 2.13 In the quaternionic case, if we impose the condition of right monogenicity, i.e.,

$$(1 \ i \ j \ k) \begin{pmatrix} \dfrac{\partial y_1}{\partial x_1} & \dfrac{\partial y_1}{\partial x_2} & \dfrac{\partial y_1}{\partial x_3} & \dfrac{\partial y_1}{\partial x_4} \\ \dfrac{\partial y_2}{\partial x_1} & \dfrac{\partial y_2}{\partial x_2} & \dfrac{\partial y_2}{\partial x_3} & \dfrac{\partial y_2}{\partial x_4} \\ \dfrac{\partial y_3}{\partial x_1} & \dfrac{\partial y_3}{\partial x_2} & \dfrac{\partial y_3}{\partial x_3} & \dfrac{\partial y_3}{\partial x_4} \\ \dfrac{\partial y_4}{\partial x_1} & \dfrac{\partial y_4}{\partial x_2} & \dfrac{\partial y_4}{\partial x_3} & \dfrac{\partial y_4}{\partial x_4} \end{pmatrix} \begin{pmatrix} 1 \\ i \\ j \\ k \end{pmatrix} = 0 \qquad (2.35)$$

with easy computations we obtain the system:

$$\frac{\partial y_1}{\partial x_1} - \frac{\partial y_2}{\partial x_2} - \frac{\partial y_3}{\partial x_3} - \frac{\partial y_4}{\partial x_4} = 0$$

$$\frac{\partial y_1}{\partial x_2} + \frac{\partial y_2}{\partial x_1} + \frac{\partial y_3}{\partial x_4} - \frac{\partial y_4}{\partial x_3} = 0$$

$$\frac{\partial y_1}{\partial x_3} - \frac{\partial y_2}{\partial x_4} + \frac{\partial y_3}{\partial x_1} + \frac{\partial y_4}{\partial x_2} = 0$$

$$\frac{\partial y_1}{\partial x_4} + \frac{\partial y_2}{\partial x_3} - \frac{\partial y_3}{\partial x_2} + \frac{\partial y_4}{\partial x_1} = 0$$

which corresponds to the well known Cauchy–Fueter conditions for the right regularity of a quaternionic function. By taking the transpose of the jacobian, with similar calculations, we obtain the Cauchy–Fueter conditions for the left regularity, i.e.,

$$\frac{\partial y_1}{\partial x_1} - \frac{\partial y_2}{\partial x_2} - \frac{\partial y_3}{\partial x_3} - \frac{\partial y_4}{\partial x_4} = 0$$

$$\frac{\partial y_1}{\partial x_2} + \frac{\partial y_2}{\partial x_1} - \frac{\partial y_3}{\partial x_4} + \frac{\partial y_4}{\partial x_3} = 0$$

$$\frac{\partial y_1}{\partial x_3} + \frac{\partial y_2}{\partial x_4} + \frac{\partial y_3}{\partial x_1} - \frac{\partial y_4}{\partial x_2} = 0 \tag{2.36}$$

$$\frac{\partial y_1}{\partial x_4} - \frac{\partial y_2}{\partial x_3} + \frac{\partial y_3}{\partial x_2} + \frac{\partial y_4}{\partial x_1} = 0$$

Remark 2.3 In various works, see [4] and the references therein, we considered functions which are Cauchy–Fueter regular with respect to $n > 1$ quaternionic variables. The regularity condition is expressed by a system consisting of $4n$ equations obtained by writing for each of the n quaternionic variables a system of the form (2.36). The analysis of this class of functions is rather complicated and is performed in [4] with algebraic methods based on the construction of a minimal free resolution of the module associated with the system. It is essential to consider $n > 1$ since in one variable the methods do not provide interesting information, basically because the matrix representing the system (2.36) is square and nondegenerate. It is however interesting to note that when considering the notion of total derivability, the algebraic study may be meaningful also when $n = 1$ since the matrix representing the system is not square. Such a study deserves to be further investigated.

Finally, we consider the case of the algebra LXXXI in [34]:

Example 2.14 In the case of the algebra LXXXI, using the basis in Example 2.7 we immediately deduce that the left monogenicity conditions are expressed by:

$$\frac{\partial y_1}{\partial x_1} - \frac{\partial y_2}{\partial x_2} = 0$$

$$\frac{\partial y_1}{\partial x_2} + \frac{\partial y_2}{\partial x_1} = 0$$

$$\frac{\partial y_1}{\partial x_3} + \frac{\partial y_2}{\partial x_4} + \frac{\partial y_3}{\partial x_1} - \frac{\partial y_4}{\partial x_2} = 0$$

$$\frac{\partial y_1}{\partial x_4} - \frac{\partial y_2}{\partial x_3} + \frac{\partial y_3}{\partial x_2} + \frac{\partial y_4}{\partial x_1} = 0.$$

These relation are, formally, similar to the conditions of left Cauchy–Fueter regularity in which in the first two equations only the derivatives with respect to x_1 and x_2 appear. However, from the point of view of the algebraic analysis the two systems are different. This is visible, for example, when taking the minimal free resolution in the case of two variables in this algebra. The first syzygies in fact are quadratic and linear whereas in the Cauchy–Fueter case there are only quadratic syzygies, see [4].

In order to relate the notions of total derivability and of monogenicity (left or right) one needs conditions on the ambient algebra and this justifies the notions of solenoidal or bisolenoidal algebra. In order to distiguish these algebras among themselves, a useful fact that may be useful in understanding the computations in Section **2** of [27] (see also Sect. 1.1) is the following

Proposition 2.1 *Let $u = (u_1, \ldots, u_n)$, $u' = (u'_1, \ldots, u'_n)$ where u and u' are bases of the algebra $\mathscr{A}$ and let $u = u'P$ where P is a nonsingular matrix. Let ξ and $\xi' = \xi P^T$ the coordinates with respect these two bases. The jacobian of ξ' with respect to ξ is $\left(\dfrac{d\xi'}{d\xi} \right) = P$.*

Proof Let $P = (p_{ij})$, then

$$\xi' = \xi P^T \qquad \Longleftrightarrow \qquad (\xi'_1, \xi'_2, \ldots, \xi'_n) = (\xi_1, \xi_2, \ldots, \xi_n) \begin{pmatrix} p_{11} & p_{21} & \cdots & p_{n1} \\ p_{12} & p_{22} & \cdots & p_{n2} \\ \cdots & & \cdots & \\ p_{1n} & p_{2n} & \cdots & p_{nn} \end{pmatrix},$$

so that

$$\xi'_1 = p_{11}\xi_1 + p_{12}\xi_2 + \ldots + p_{1n}\xi_n$$
$$\xi'_2 = p_{21}\xi_1 + p_{22}\xi_2 + \ldots + p_{2n}\xi_n$$
$$\ldots = \ldots$$
$$\xi'_n = p_{n1}\xi_1 + p_{n2}\xi_2 + \ldots + p_{nn}\xi_n.$$

Thus $\dfrac{\partial \xi'_i}{\partial \xi_j} = p_{ij}$ and the statement follows.   □

Let us recall the following formulas:

$$u_i u P P_{-1} u_{-1} = 0, \qquad i = 1, \ldots, n \tag{2.10}$$

$$u P P_{-1} u_i u_{-1} = 0, \qquad i = 1, \ldots, n. \tag{2.11}$$

$$zu P P_{-1} u_{-1} = 0, \tag{2.12}$$

$$u P P_{-1} zu_{-1} = 0; \tag{2.13}$$

$$u P P_{-1} u_{-1} = 0. \tag{2.14}$$

Definition 2.8 An algebra with unit is solenoidal if its bases satisfy (2.14) with P non singular. It is bisolenoidal if it has bases satisfying (2.11) with P non singular.

Theorem 2.2 *In an algebra with unit we have*
$(2.10) \Longleftrightarrow (2.12) \Longleftrightarrow (2.14)$
and
$(2.11) \Longleftrightarrow (2.13) \Longrightarrow (2.14).$
Thus, if an algebra is bisolenoidal it is also solenoidal, but not viceversa.

The results in Sect. 2.1 can be summarized as follows:

Theorem 2.3 *In an algebra with unit, functions right totally derivable are right monogenic if and only if (2.14) holds while they are left monogenic if and only if the relations (2.11) hold. In this latter case, the functions are also right monogenic.*

Proof Assume that a function $y(x)$ is right totally derivable, then the condition of being right monogenic, with respect to a basis obtained via the change of basis given by the nonsingular matrix P, is expressed by (2.10) which hold if and only if (2.14) holds. The condition of being left monogenic, with respect to a basis obtained via the change of basis given by the nonsingular matrix P, is expressed by (2.11). Since (2.11) imply (2.14), the functions are automatically also right monogenic. $\quad\square$

Remark 2.4 We believe that in the original manuscript there is a typo and the sentence
    (2.14) and (2.11) with P nonsingular, are necessary and sufficient conditions for right totally derivable functions in an algebra $\mathscr{A}$ to be right or left monogenic (with respect to a suitable basis) should be instead
    (2.14) or (2.11) with P nonsingular, are necessary and sufficient conditions for right totally derivable functions in an algebra $\mathscr{A}$ to be right or left monogenic (with respect to a suitable basis).
    The amended sentence is also in accordance to the review of paper written by B. Crabtree, see [5].

To illustrate these ideas we consider some examples in the case of algebras of low order.

Example 2.15 Let us consider the algebra of dual numbers, see Example 2.2. Setting $PP^T = A = (a_{jk})$ we compute (2.14):

$$(u_1 \ u_2) \begin{pmatrix} a_{11} & a_{12} \\ a_{12} & a_{22} \end{pmatrix} \begin{pmatrix} u_1 \\ u_2 \end{pmatrix} = 0.$$

We obtain $a_{11}u_1 + a_{12}u_2 = 0$ which immediately gives $a_{11} = a_{12} = 0$ and PP^T and so P are singular. From this fact one immediately deduces that the algebra of dual numbers is neither solenoidal, nor bisolenoidal. Note that (2.11) for $i = 1$ reduces to the condition above and for $i = 2$ it is an identity.

Example 2.16 To illustrate the reasoning in Sect. 2.2, we develop the computations in the case of ternions. We need to compute the elements a_{ik} of the matrix A using (2.14). We have:

$$(u_1 \; u_2 \; u_3) \begin{pmatrix} a_{11} & a_{12} & a_{13} \\ a_{12} & a_{22} & a_{23} \\ a_{13} & a_{23} & a_{33} \end{pmatrix} \begin{pmatrix} u_1 \\ u_2 \\ u_3 \end{pmatrix} = 0,$$

which leads to

$$a_{11}u_1 + a_{22}u_2 + a_{23}u_3 + a_{13}u_3 = 0.$$

We deduce that $a_{11} = a_{22} = 0$ and $a_{13} + a_{23} = 0$ so that

$$A = \begin{pmatrix} 0 & a_{12} & a_{13} \\ a_{12} & 0 & -a_{13} \\ a_{13} & -a_{13} & a_{33} \end{pmatrix}.$$

Although A is nonsingular, it is not positive as one may notice taking the principal minor, i.e. the determinant of

$$\begin{pmatrix} 0 & a_{12} \\ a_{12} & 0 \end{pmatrix}$$

which has negative value, so A cannot be of the form PP^T. We conclude that the algebra of ternions is not solenoidal over the real.

We note that if we would have considered the ternions over the complex field, then the complexified ternion algebra is solenoidal since the only condition required on A is of being nonsingular (see Sect. 1.2, paragraph **4**). We now consider the conditions in (2.11):

$$(u_1 \; u_2 \; u_3) \begin{pmatrix} a_{11} & a_{12} & a_{13} \\ a_{12} & a_{22} & a_{23} \\ a_{13} & a_{23} & a_{33} \end{pmatrix} \begin{pmatrix} u_i u_1 \\ u_i u_2 \\ u_i u_3 \end{pmatrix} = 0, \qquad i = 1, 2, 3.$$

Taking into account the relations satisfied by u_1, u_2, u_3 we obtain the following system:

$$a_{11}u_1^2 + a_{13}u_1u_3 = 0$$

$$a_{12}u_1u_2 + a_{22}u_2^2 + a_{23}u_3u_2 = 0$$

$$a_{12}u_1u_3 + a_{22}u_2u_3 + a_{23}u_3^2 = 0$$

which yields to

$$a_{11}u_1 + a_{13}u_3 = 0$$
$$a_{22}u_2 + a_{23}u_3 = 0$$
$$a_{12}u_3 = 0.$$

We deduce that $a_{11} = a_{13} = a_{22} = a_{23} = a_{12} = 0$, so that A, and so P, are singular. We conclude that the algebra of ternions is not bisolenoidal.

As algebra of fourth order, we consider the algebra of bicomplex numbers.

Example 2.17 The condition (2.14) is

$$(1 \; i \; j \; k) \begin{pmatrix} a_{11} & a_{12} & a_{13} & a_{14} \\ a_{12} & a_{22} & a_{23} & a_{24} \\ a_{13} & a_{23} & a_{33} & a_{34} \\ a_{14} & a_{24} & a_{34} & a_{44} \end{pmatrix} \begin{pmatrix} 1 \\ i \\ j \\ k \end{pmatrix} = 0.$$

These conditions translate into

$$a_{11} - a_{22} - a_{33} + a_{44} = 0$$
$$a_{12} - a_{34} = 0$$
$$a_{13} - a_{24} = 0$$
$$a_{14} + a_{23} = 0,$$

which shows, with some more computations to show the positivity, that the bicomplex algebra is solenoidal. Since the algebra is commutative, the conditions (2.11) follows form the previous ones and thus the algebra is also bisolenoidal. Thus right total derivability implies monogenicity (right and left) as shown by Example 2.34. Note that one can take $P = A = I$, the identity matrix, in the above computations and this is in accordance with the fact that (right) total derivability imply (right) monogenicity with respect to the same basis.

Example 2.18 As already noticed in Sect. 1.1 the algebra of quaternions is solenoidal. This can also be verified by direct computations which lead to

$$a_{11} - a_{22} - a_{33} - a_{44} = 0$$
$$a_{12} = 0$$
$$a_{13} = 0$$
$$a_{14} = 0,$$

and thus to the matrix

$$\begin{pmatrix} a_{22} + a_{33} + a_{44} & 0 & 0 & 0 \\ 0 & a_{22} & a_{23} & a_{24} \\ 0 & a_{23} & a_{33} & a_{34} \\ 0 & a_{24} & a_{34} & a_{44} \end{pmatrix}.$$

It is possible to choose the matrix A such that it is nonsingular.

References

1. Albert, A.A.: Structure of Algebras, vol. XXIV. American Mathematical Society Colloquium Publications, New York (1939)
2. Albert, A.A.: Modern Higher Algebra. The University of Chicago Science Series, Chicago (1937)
3. Brackx, F., Delanghe, R., Sommen, F.: Clifford analysis. In: Research Notes in Mathematics, vol. 76. Pitman (Advanced Publishing Program), Boston (1982)
4. Colombo, F., Sabadini, I., Sommen, F., Struppa, D.C.: Analysis of Dirac systems and computational algebra. In: Progress in Mathematical Physics, vol. 39 (Birkhäuser, Boston (2004)
5. Crabtree, B.: Review MR0067851, MathSciNet, Mathematical Reviews
6. Fueter, R.: Über Funktionen einer Quaternionenvariablen. Atti Congresso Bologna **2**, 145 (1930)
7. Fueter, R.: Analytische Funktionen einer Quaternionenvariablen. Comment. Math. Helv. **4**, 9–20 (1932)
8. Fueter, R.: Die Funktionentheorie der Differentialgleichungen $\Delta u = 0$ und $\Delta \Delta u = 0$ mit vier reellen Variablen. Comment. Math. Helv. **7**, 307–330 (1934–1935)
9. Fueter, R.: Über die Funktionentheorie in einer hypercomplexen Algebra. Elem. Math. III **5**, 89–104 (1948)
10. Gürlebeck, K., Malonek, H.R.: A hypercomplex derivative of monogenic functions in R^{n+1} and its applications. Compl. Var. Theory Appl. **39**, 199–392 (1999)
11. Haefeli, H.: Hypercomplex Differential. Comment. Math. Helv. **20**, 382–420 (1947)
12. Iftimie, V.: Fonctions Hypercomplexes. Bull. Math. Soc. Sci. Math. R. S. Roumanie **9**, 279–332 (1965)
13. Iwanenko, D., Nikolsky, K.: Über den Zusammenhang zwischen den Cauchy–Riemannschen und Diracschen Differentialgleichungen. Z. Physik **63**, 129–137 (1930)
14. Ketchum, P.W.: Analytic functions of hypercomplex variables. Trans. Am. Math. Soc. **30**, 641–667 (1928)
15. Kriszten, A.: Hypercomplexe und pseudo-analytische Funktionen. Comment. Math. Helv. **26**, 6–35 (1952)
16. Krylov, N.M.: Dokl. AN SSSR **55**(9), 799–800 (1947)
17. Lanczos, C.: The Functional Theoretical Relationships of the Maxwell Aether2Equations a Contribution to the Theory of Relativity and Electrons, Dissertation (1919)
18. Lanczos, C.: Collected Published Papers with Commentaries, vol. IVI. North Carolina State University, College of Physical and Mathematical Sciences, Raleigh (1998)
19. Luna-Elizarraras, M.E., Shapiro, M.: A survey on the (hyper)-derivatives in complex, quaternionic and Clifford analysis. Milan J. Math. **79**, 521–542 (2011)
20. Luna-Elizarras, M.E., Shapiro, M., Struppa, D.C., Vajiac, A.: Bicomplex Holomorphic Functions. Frontiers in Mathematics, Birkhäuser (2015)

21. Malonek, H.R.: A new hypercomplex structure of the Euclidean space R^{n+1} and the concept of hypercomplex derivative. Compl. Var. Theory Appl. **14**, 25–33 (1990)
22. Meilikhson, A.S.: Dokl. AN SSSR **59**(3), 431–434 (1948)
23. Moisil, G.C.: Sur les quaternions monogènes. Bull. Sci. Math. (Paris) **LV**, 168–174 (1931)
24. Nef, W.: Ueber eine Verallegemeinnerung des Satzes von Fatou für Potentialfunktionen. Comme. Math. Helv. **13**, 215–241 (1943–1944)
25. Price, G.B.: An introduction to multicomplex spaces and functions. In: Monographs and Textbooks in Pure and Applied Mathematics, vol. 140. Marcel Dekker, New York (1991)
26. Rizza, G.B.: Sulle funzioni analitiche nelle algebre ipercomplesse. Pont. Acad. Sci. Comment. **14**, 169–194 (1950)
27. Sce, M.: Monogeneità e totale derivabilità nelle algebre reali e complesse. I. Atti Accad. Naz. Lincei. Rend. Cl. Sci. Fis. Mat. Nat. (8) **16**, 30–35 (1954, Italian)
28. Sce, M.: Monogeneità e totale derivabilità nelle algebre reali e complesse. II. Atti Accad. Naz. Lincei. Rend. Cl. Sci. Fis. Mat. Nat. (8) **16**, 188–193 (1954, Italian)
29. Sce, M.: Monogeneità e totale derivabilità nelle algebre reali e complesse. III. Atti Accad. Naz. Lincei. Rend. Cl. Sci. Fis. Mat. Nat. (8) **16**, 321–325 (1954, Italian)
30. Scorza, G.: Sopra un teorema fondamentale della teoria delle algebre. Rend. Acc. Lincei (6) **20**, 65–72 (1934)
31. Scorza, G.: Le algebre per ognuna delle quali la sottoalgebra eccezionale potenziale. Atti R. Acc. Scienze di Torino **70**(11), 26–45 (1934–1935)
32. Scorza, G.: Le algebre doppie. Rend. Acc. Napoli (3) **28**, 65–79 (1922)
33. Scorza, G.: Le algebre del 3° ordine. Acc. Napoli (2) **20**, n.13 (1935)
34. Scorza, G.: Le algebre del 4° ordine. Acc. Napoli (2) **20**, n.14 (1935)
35. Scorza Dragoni, G.: Sulle funzioni olomorfe di una variabile bicomplessa. Mem. Acc. d'Italia **5**, 597–665 (1934)
36. Segre, B.: Forme differenziali e loro integrali. Roma (1951)
37. Sobrero, L.: Algebra delle funzioni ipercomplesse e una sua applicazione alla teoria matematica dell'elasticità. Mem. R. Acc. d'Italia **6**, 1–64 (1935)
38. Spampinato, N.: Sulle funzioni totalmente derivabili in un'algebra reale o complessa dotata di modulo. Rend. Lincei **21**, 621–625 (1935)
39. Subdery, A.: Quaternionic Analysis. In: Mathematical Proceedings of the Cambridge Philosophical Society, vol. 85, 199–225 (1979)
40. Ward, J.A.: A theory of analytic functions in linear associative algebras. Duke Math J. **7**, 233–248 (1940)

Chapter 3
On Systems of Partial Differential Equations Related to Real Algebras

This chapter contains the translation of the paper:

M. Sce, *Sui sistemi di equazioni differenziali a derivate parziali inerenti alle algebre reali*, (Italian) Atti Accad. Naz. Lincei. Rend. Cl. Sci. Fis. Mat. Nat. (8) **18** (1955), 32–38.

Article by Michele Sce, presented during the meeting of 11 December 1954 by B. Segre, member of the Academy.

In this Note, after some preliminaries in algebra and analysis, we classify systems of partial differential equations which give monogenicity conditions in algebras. These systems are elliptic for primitive algebras, parabolic with algebras with radical; the proof of this latter fact is based on a characterization, maybe unknown, of semisimple algebras via their determinant. Among hyperbolic systems, we highlight the one obtained from a regular algebra, and which has as characteristic hypersurface the cone of zero divisors of the algebra itself.

From these systems we deduce some partial differential equations of order equal to the order of the algebra which are of the same type of those satisfied by all monogenic functions. I hope that a further study of these equations would eventually lead me to solve, at least for monogenic functions in regular algebras, some problems analogous to the Cauchy problem.

1. Let $\mathscr{A}$ be an associative algebra over the real, with unit, and let $U = (u_1, \ldots, u_n)$ be a basis. The algebras of matrices $\mathscr{A}'$ ($\mathscr{A}''$) which gives the first (second) regular representation of $\mathscr{A}$ are such that for every x in $\mathscr{A}$ and X' (X'') in $\mathscr{A}'$ ($\mathscr{A}''$) one has

$$xu_{-1} = X'_{-1}u_{-1} \qquad (ux = uX''_{-1}), \qquad (3.1)$$

the determinant of X', (X'') is also called *left (right) determinant* of x.[1]

The element $y = y_1 u_1 + \cdots + y_n u_n = \eta u_{-1}$ in $\mathscr{A}$ is said to be left (right) monogenic function of $x = x_1 u_1 + \cdots + x_n u_n$ if y_i are functions derivable with respect to x_k and such that

$$\sum_{i,k} \frac{\partial y_i}{\partial x_k} u_k u_i = 0 \qquad \left(\sum_{i,k} \frac{\partial y_i}{\partial x_k} u_i u_k = 0 \right) .[2]$$

By setting

$$u_k u_i = \sum_j c_{ki}^j u_j, \tag{3.2}$$

one gets a system of n linear differential equations of the first order:

$$\sum_{i,k} \frac{\partial y_i}{\partial x_k} c_{ki}^j = 0 \qquad (j = 1, 2, \ldots, n), \tag{3.3}$$

whose coefficients are the constants of multiplication in the algebra. By setting [Editors' Note: $\|a_{ij}\|$ denotes the matrix with entries a_{ij}]

$$\Omega = \left\| \sum_k c_{ki}^j \frac{\partial}{\partial x_k} \right\| = \|\gamma_{ij}\| \qquad (i, j = 1, 2, \ldots, n),$$

(3.3) can be written in a more compact form as:

$$\Omega \eta_{-1} = 0; \tag{3.4}$$

the determinant $|C|$ of the matrix

$$C = \left\| \sum_k c_{ki}^j z_k \right\| = \sum_k C_k z_k, \tag{3.5}$$

$C_k = \|c_{ki}^j\|$ (for $i, j = 1, \ldots, n$), obtained from Ω by substituting z_k instead of $\dfrac{\partial}{\partial x_k}$, is called *characteristic form* of the system. The system itself is called *elliptic* or *parabolic* if the *characteristic equation*

$$|C| = f(z_1, \ldots, z_n) = 0 \tag{3.6}$$

---

[1] See G. Scorza, *Corpi numerici e algebre*, Messina (1921), Part II, n. 184 and 185.

[2] See B. Segre, *Forme differenziali e loro integrali*, (Roma, 1951), Ch. IV, n. 90. The indices of the infinite sums, unless otherwise stated, always run from 1 to n. Later we will always refer to left monogenicity, since analogous results can be obtained in a similar way for right monogenicity.

does not have real solutions different from the trivial solution $(0, \ldots, 0)$ or, by means of a change of variables, it can be reduced to an equation depending on less than n variables;[3] in the other cases the system is called *hyperbolic* and, in particular, *totally hyperbolic* if the matrix C is diagonalizable and its characteristic polynomial has real roots whatever are the z_k's.[4]

Finally, a hypersurface $\varphi(x_1, \ldots, x_n) = 0$ in the Euclidean space of n-tuples $(x_1, \ldots, x_n)$ is called *characteristic* if the directional cosines of its normal

$$\frac{\partial \varphi_1}{\partial x_1}, \ldots, \frac{\partial \varphi_n}{\partial x_n}$$

satisfy the characteristic equation (3.6).[5]

2. Let us multiply the matrix (3.5) on the right by the n-vector u_{-1}; in force of (3.2), one has:

$$Cu_{-1} = \begin{pmatrix} \sum_{j,k} c_{k1}^j z_k u_j \\ \vdots \\ \sum_{j,k} c_{kn}^j z_k u_j \end{pmatrix} = \begin{pmatrix} \sum_k z_k u_k u_1 \\ \vdots \\ \sum_k z_k u_k u_n \end{pmatrix}$$

and, setting $z = z_1 u_1 + \cdots + z_n u_n$, we deduce

$$zu_{-1} = Cu_{-1}.$$

A comparison with (3.1) shows that C_{-1} is the matrix corresponding to z in the first regular representation of $\mathscr{A}$;[6] thus *the characteristic form of system* (3.3) *coincides with the left determinant of an element in* $\mathscr{A}$.

Since the determinants of elements in $\mathscr{A}$ are invariant with respect to change of basis in the algebra, one has also that *the characteristic form of the system expressing the monogenicity is invariant with respect to changes of basis in the algebra*.[7]

---

[3] See R. Courant, D. Hilbert, *Methoden der Mathematischen Physik*, Band II (Berlin, 1937), Kap. III, § 4, n.2.

[4] A definition equivalent to ours can be found, for systems of quasi-linear equations, in R. Courant, K. O. Friedrichs, *Supersonic flow and shock waves*, Interscience Publishers, Inc., New York, N. Y., 1948, Chapt. II, n. 32; sometimes, as in R. Courant, P. Lax, *On nonlinear partial differential equations with two independent variables*, Comm. Pure Appl. Math., **2** (1949), 255–273, pp. 255–273, n.2, totally hyperbolic systems are called *hyperbolic*.

[5] See Courant-Friedrichs, cited in [4].

[6] In particular, Ω_{-1} corresponds to $\omega = \sum_k \dfrac{\partial}{\partial x_k} u_k$ considered as an element in $\mathscr{A}$ and the transpose of the C_k's in (3.5) correspond to the units in $\mathscr{A}$.

[7] On the contrary, monogenicity conditions depend on the basis of the algebra; see M. Sce, *Monogeneità e totale derivabilità nelle agebre reali e complesse*, Atti Accad. Naz. Lincei. Rend. Cl. Sci. Fis. Mat. Nat., (8) **16** (1954), 30–35, Nota I, n. 1.

Primitive algebras have no zero divisors and so the determinants of their nonzero elements are always nonzero; and conversely. Based on the preceding arguments and on n. 1, this is equivalent to claim that *system* (3.3) *is elliptic for primitive algebras and only for those.*

3. We will prove that *the system* (3.3) *is parabolic for algebras not semi-simple and only for them,* by proving that *an algebra is semi-simple if and only if, with respect to any basis, the determinant of the elements depends on all their coordinates.*

Let us assume the $\mathscr{A}$ is semi-simple and that the determinant of the elements does not depend on all their coordinates. $\mathscr{A}$ is direct sum of simple algebras $\mathscr{A}_i$ and so the determinant of the elements in $\mathscr{A}$ is the product of the determinants of the elements in $\mathscr{A}_i$; thus some algebra $\mathscr{A}_i$ is such that the determinant of its elements $x = \sum_i x_i u_i$, $y = \sum_i y_i u_i$ does not depend, for example, on the coefficients of the units $u_m, \ldots, u_n$ but it depends on the coefficients of all the other units. Then also the determinant of

$$xy = \sum_{i,k=1}^{m-1} x_i y_k u_i u_k + \sum_{i=1,\ldots n, k=m,\ldots n} (x_i y_k u_i u_k + x_k y_i u_k u_i) + \sum_{i=m}^{n} x_i y_i u_i^2$$

does not depend on $x_m, \ldots, x_n$; $y_m, \ldots, y_n$; it turns out that all the products of units appearing in the second sum can be expressible as linear combinations of $u_m, \ldots, u_n$, namely

$$u_i u_k = \sum_{j=m}^{n} c_{ik}^j u_j \qquad (i = 1, \ldots, n; \; k = m, \ldots, n) \qquad (3.7)$$

must hold. The relations (3.7) allow to say that the set having basis $u_m, \ldots, u_n$ is a proper ideal of $\mathscr{A}_i$; this contradicts the assumption that $\mathscr{A}_i$ is simple and shows the necessity of the condition.

Let now $\mathscr{A}$ an algebra not semi-simple, that is, it has a nonzero radical $\mathscr{R}$. Let $U_1, \ldots, U_m, \ldots, U_n$, $(1 < m < n)$ be a basis for $\mathscr{A}'$ such that $U_m, \ldots, U_n$ is a basis for the image $\mathscr{R}'$ of $\mathscr{R}$ in the first regular representation of $\mathscr{A}$; then the trace of any element $X' = \sum_i x_i U_i$ in $\mathscr{A}'$ does not depend on $x_m, \ldots, x_n$ since the matrices $U_m, \ldots, U_n$—which are elements of $\mathscr{R}'$ are nilpotent. Since

$$X'^2 = \sum_{i,k=1}^{m-1} x_i x_k U_i U_k + \sum_{i=1,\ldots,n, \; k=m,\ldots,n} x_i x_k (U_i U_k + U_k U_i) - \sum_{i=m}^{n} x_i^2 u_i^2,$$

only the coefficients of the first sum—among which $x_m, \ldots, x_n$ do not appear—can give a nonzero contribution to the trace of X'^2 in fact $\mathscr{R}'$ is an ideal of $\mathscr{A}'$ and so a relation similar to (3.7) holds. Reasoning in this way we can prove that the traces of $X', X'^2, \ldots, X'^n$ do not depend on $x_m, \ldots, x_n$. Using the recurrence formulas—

which can be easily obtained[8]—expressing the coefficients of the characteristic equation of a matrix via the traces of a matrix and its powers, one gets that each coefficient of the characteristic equation of X' does not depend on $x_m, \ldots, x_n$. This is true, in particular, for the determinant of x and so the theorem is proved.

4. *The system* (3.3) *is completely hyperbolic for the algebras of n-real numbers and only for them.*

Let $u_1, \ldots, u_n$, $u_i^2 = u_i$, $u_i u_k = 0$, $(i \neq k)$, be the basis of n-real numbers; the matrix (3.5) turns out to be real and diagonal,so that the system (3.3) is totally hyperbolic.

Conversely, let us assume that the system (3.3) is totally hyperbolic, that is, C is diagonalizable and its characteristic roots are real. If C_i is diagonal and all the z_k except z_i, z_j are zero, the matrix $z_i C_i + z_j C_j$ is diagonal only if C_j is diagonal; it follows that, when i and j vary, a matrix which reduces a C_k to the diagonal form must reduce to the diagonal form all the other C_k. Thus the matrices C_k are pairwise commuting;[9] and since their transpose correspond to the units of $\mathscr{A}$ in the first regular representation, $\mathscr{A}$ is commutative. On the other hand, since the system (3.3) is not parabolic, $\mathscr{A}$ is semi-simple (n. **3**); a semi-simple algebra which is commutative is direct sum of the real field and of the algebra of complex numbers, and both can be repeated a certain number of times.[10] However, if $\mathscr{A}$ would have as a component the algebra of complex numbers, some C_k would have complex characteristic roots; thus $\mathscr{A}$ cannot be anything but the direct sum of the real field taken n times.

5. Let $e_{11}, e_{12}, \ldots, e_{nn}$ with $e_{ij}e_{jk} = e_{ik}$, $e_{ij}e_{lk} = 0$, $(i \neq l)$ the basis of a regular algebra $\mathscr{M}_n$ of order n^2; the matrices of order n^2 elements of $\mathscr{M}'_n$ are direct sums of n matrices equal to $\|x_{ik}\|$, of order n, in $\mathscr{M}_n$. Thus $\varphi(x_{ik}) = 0$ is a characteristic hypersurface of system (3.3) if:

$$\frac{\partial \varphi}{\partial x_{ik}} = 0. \tag{3.8}$$

Since $\dfrac{\partial |x_{ik}|}{\partial x_{ik}}$ is the adjoint of X_{ik} in $\|x_{ik}\|$, one has

$$\left| \frac{\partial |x_{ik}|}{\partial x_{ik}} \right| = |x_{ik}|^{n-1}$$

and the cone $|x_{ik}| = 0$ of zero divisors of $\mathscr{M}_n$ satisfies (3.8); thus *the cone of zero divisors of a regular algebra $\mathscr{M}_n$ counted $n - 1$ times, is a characteristic hypersurface of system* (3.3).

---

[8]When a matrix is in canonic form, these formulas reduce to those of symmetric functions.

[9]See M. Sce, *Su alcune proprietà delle matrici permutabili e diagonalizzabili*, Rivista di Parma, vol. 1, (1950), pp. 363–374, n.5.

[10]See Scorza cited in [(1)], part II, n. 292.

Let us now consider the algebra $\mathscr{A}$ of order $2n^2$ direct product of the algebra of complex numbers and the regular algebra $\mathscr{M}_n$. The elements of $\mathscr{A}'$ with respect to $e_{ik}, i e_{ik}$ are matrices of order $2n^2$ direct sums of n matrices of order $2n$,

$$X = \begin{pmatrix} A & -B \\ B & A \end{pmatrix},$$

where A, B are arbitrary elements of $\mathscr{M}_n$. Denoting again by $X_{j\ell}$ the adjoint of $x_{j\ell}$ (where $j, \ell = 1, 2, \ldots, 2n$, a simple computation shows that—for $i, k = 1, \ldots, n$— one has:

$$\left| \frac{\partial |x_{j\ell}|}{\partial x_{j\ell}} \right| = \left| \begin{matrix} X_{ik} + X_{n+i,n+k} & X_{i,n+k} - X_{n+i,k} \\ X_{n+i,k} + X_{i,n+k} & X_{n+i,n+k} + X_{i,k} \end{matrix} \right| = |X_{j\ell}|;$$

thus the equation $|x_{j\ell}|^{2n^2-1} = 0$—in the complex case—represents a characteristic hypersurface of system (3.3). However, it should be noted that, in the real field, $|x_{j\ell}| = 0$ represents a cone (with vertex at the origin) of dimension $2(n^2-1)$ which certainly does not give a characteristic hypersurface.

Analogous remarks can be made in the case of an algebra direct product of the algebra of quaternions with a regular algebra.

6. Let $A = \|a_{ik}\|$ be a matrix of order n whose elements belong to an integral domain $\mathscr{D}$ with unit, but which is not a principal ideal ring. If A has maximal rank, its first column cannot be zero; thus, multiplying A on the left by a suitable matrix, we can assume that the element $(1, 1)$ is nonzero. Let us still denote by A the matrix obtained in this way, and let us multiply it by the nonsingular matrix

$$P = \begin{pmatrix} 1 & 0 & 0 & \ldots \\ -a_{21} & a_{11} & 0 & \ldots \\ -a_{31} & 0 & a_{11} & \ldots \\ \vdots & \vdots & \vdots & \ldots \end{pmatrix};$$

one obtains

$$PA = \begin{pmatrix} a_{11} & a_{12} & a_{13} & \ldots \\ 0 & a_{11}a_{22} - a_{12}a_{21} & a_{11}a_{23} - a_{13}a_{21} & \ldots \\ 0 & a_{11}a_{32} - a_{12}a_{31} & a_{11}a_{33} - a_{13}a_{31} & \ldots \\ \vdots & \vdots & \vdots & \ldots \end{pmatrix} = \begin{pmatrix} a_{11} & a \\ 0 & A_1 \end{pmatrix}. \tag{3.9}$$

Since also the first column of A_1 must be nonzero, we can act on A_1 as we did on A; iterating the procedure, we show that *every nonsingular matrix in $\mathscr{D}$ can be reduced in triangular form T, multiplying it on the left by a suitable nonsingular matrix.*

As it can be seen from (3.9), the first two elements on the principal diagonal of T are two minors of order 1 and 2 respectively, the first one contained in the second; reasoning by induction one finds that—by selecting a sequence of nonzero minors $\alpha_1, \alpha_2, \ldots, \alpha_n = |A|$ of all the orders from 1 to n, each of which contained in the following (complete chain)—*the elements on the principal diagonal of T are*

$$\alpha_1, \ \alpha_2, \ \alpha_1\alpha_3, \ \alpha_1^2\alpha_2\alpha_4, \ \ldots, \ \alpha_1^{n-2}\alpha_2^{n-3}\cdots\alpha_{n-2}\alpha_n.^{11} \tag{3.10}$$

The degree of the i-th element in the sequence (3.10) in the elements of A is clearly $\sum_{k=1}^{i-2} k2^{i-2-k} + i$; since

$$\sum_{k=i-1}^{\infty} k2^{i-2-k} = i,$$

we can conclude that *the i-th element of the sequence (3.10) is of degree 2^{i-1} in the elements of A.*

Let us consider the extension $\mathscr{F}$ of the field of the real numbers by means of the operators $\dfrac{\partial}{\partial x_1}, \ldots, \dfrac{\partial}{\partial x_n}$; defining formally, as usual, the operations of sums and product, $\mathscr{F}$ turns out to be a ring with unit. The set of operators $\mathscr{F}$ has no zero divisors; thus, if we assume that all the functions to which we apply elements of $\mathscr{F}$ have finite derivatives, continuous up to the order m, we can say that all the elements of order not greater than m behave like elements in an integral domain.

Thus, if we suppose that the y_i elements of η in (3.4) possess finite derivatives, continuous up to order 2^{n-1}, we can apply to the matrix Ω the considerations made above, this gives for the element y_n of η a partial differential equation of order at most 2^{n-1}. An analogous result can be obtained for the other y_i; but, in general, the equations obtained for the various y_i are different. To obtain a differential equation satisfied by all the y_i's, one needs to multiply between them the n complete chains, assuming that the operators which are their elements commute (namely, that the functions have derivatives finite and continuous up to the order of the equation);

---

[11] Obviously for particular matrices one can obtain much more; for example, if

$$A = \begin{pmatrix} A' & A''' \\ -A''_{-1} & A'' \end{pmatrix}$$

is a symmetric matrix of the fourth order one has:

$$\begin{pmatrix} A'J & -A''J \\ -A''_{-1}J & A'J \end{pmatrix}\begin{pmatrix} A' & A'' \\ -A''_{-1} & A''' \end{pmatrix} = \begin{pmatrix} D & 0 \\ 0 & D \end{pmatrix}, \text{ with } J = \begin{pmatrix} 0 & -1 \\ 1 & 0 \end{pmatrix}, \ D = \begin{pmatrix} 0 & d \\ -d & 0 \end{pmatrix}$$

where d is the pfaffian of A. Another example is given by the elemnts in the algebra which is the first regular representation of quaternions; these matrices multiplied by the transposed give the scalar matrix $|q|I_4$ (where $|q|$ is the norm of the quaternion q).

taking into account that the n-chains have all the same last element, which does not have to be necessarily repeated in the product, one has that *the single components of a monogenic function satisfy a same partial differential equation of order at most* $n(2^{n-1} - (n-1))$.

7. The procedure illustrated in the preceding section, is maybe of some interest; however, an equation satisfied by all the y_i's can be easily be obtained by multiplying (3.4) on the left by the matrix Ω^* adjoint of Ω. So we have that *the components of a monogenic function satisfy an equation of order at most n.*[12]

The characteristic form of the equation coincides with the characteristic form of the system; thus *the equations satisfied by the components of the monogenic functions are of the same kind of system* (3.3) *and the characteristic hypersurfaces of the system are the same of those of the equation.*

Since, as we have seen in n. **5**, the elements of the first regular representation of an algebra simple of order kn^2 are composed by n matrices and Ω, as we observed in the note 6, can be considered an element of the first regular representation, one has that *the single components of a monogenic function in a simple algebra of order kn^2 satisfy an equation of order at most kn.*

It is easy to extend the result to semisimple algebras.

3.1   Comments and Historical Remarks

In the paper translated in this chapter, Sce discusses the problem of characterising an algebra according to the properties of the system satisfied by the monogenic functions in that algebra see [8]. In his papers, he always has a special taste for algebraic questions, see also his paper [9], and in fact inspired by this problem, he also proves a new property of semi-simple algebras. Moreover, he also shows that each function, which is a component of a monogenic function, satisfies a suitable system of differential equations with order equal to the order of the algebra. Also this paper is an interesting combination of properties exquisitely algebraic in nature and analytical properties of functions.

Most of the properties of the algebras considered in this chapter are given in Chap. 2. The reader may refer to the books of Albert [1, 2] and of Scorza [10] which were also used by Sce.

Remark 3.1 In the paper we consider in this chapter, Sce makes use of the Italian term *"algebra primitiva"*, i.e., *primitive algebra*, that we keep in the translation.

---

[12]Sometimes, by multiplying Ω on the left by matrices different from Ω^*, we obtain equations of lesser order; for example, in the case of quaternions multiplying Ω by its transpose one obtains the scalar matrix ΔI_n (Δ is the laplacian in four variables).

In more modern terms, one should translate the term as *division algebra*. In fact, as one can read in various old sources, see e.g. [11–13], a primitive algebra is an algebra which does not contain zero divisors. Such an algebra is semi-simple and also simple.

Another important definition is the following:

Definition 3.1 Let $\mathscr{A}_1, \ldots, \mathscr{A}_m$ be algebras over a field F. We say that $\mathscr{A}$ is the direct sum of $\mathscr{A}_1, \ldots, \mathscr{A}_m$ and we write

$$\mathscr{A} = \mathscr{A}_1 \oplus \ldots \oplus \mathscr{A}_m$$

if $v_i \, v_j = 0$ for $v_i \in \mathscr{A}_i$, $i \neq j$ and the order of $\mathscr{A}$ is the sum of the orders of $\mathscr{A}_i$, $i = 1, \ldots, m$.

Example 3.1 An instance of algebra which is a direct sum used in this chapter is the algebra of n-real numbers $\mathbb{R}^n = \mathbb{R} \oplus \cdots \oplus \mathbb{R}$ in which all the various copies of $\mathbb{R}$ are generated by u_i (the unit of $\mathbb{R}$) for $i = 1, \ldots, n$, and are such that $u_i u_j = 0$ when $i \neq j$.

An algebra is said to be decomposable if it can be written as sum of its (nontrivial) subalgebras, indecomposable if this is not possible.

These notions about the algebras are useful to characterise the systems of differential equations associated with the various notions of monogenicity. To provide concrete examples, we consider the monogenicity conditions in Chap. 1, for increasing dimension of the algebras considered.

Example 3.2 Let us start by considering second order algebras. We begin with the real algebra of complex numbers, which is clearly a division algebra. Then

$$C = \begin{pmatrix} x_1 & -x_2 \\ x_2 & x_1 \end{pmatrix},$$

so that $|C| = x_1^2 + x_2^2$ and the system is, as it is well known, elliptic.

Again in dimension 2, we can consider the algebra of dual numbers and the left monogenicity condition, see (2.32), which are associated with the matrix

$$C = \begin{pmatrix} x_1 & 0 \\ x_2 & x_1 \end{pmatrix},$$

so that $|C| = x_1^2$ and the system is parabolic, which is consistent with the fact that the algebra of dual numbers is not semi-simple.

Example 3.3 In the case of hyperbolic numbers, the monogenicity is expressed by the matrix

$$C = \begin{pmatrix} x_1 & x_2 \\ x_2 & x_1 \end{pmatrix}.$$

We have that $|C| = x_1^2 - x_2^2$. Moreover, the matrix is real symmetric with eigenvalues $x_1 \pm x_2$, thus the system is completely hyperbolic. According to the result proven in n. **4** the algebra of hyperbolic numbers is, up to a suitable isomorphism, the algebra $\mathbb{R}^2$. To see this fact, let us set

$$\mathbf{e} = \frac{u_1 + u_2}{2}, \qquad \mathbf{e}^\dagger = \frac{u_1 - u_2}{2}.$$

Then $\mathbf{e}$, $\mathbf{e}^\dagger$ are two idempotents such that $\mathbf{e}\,\mathbf{e}^\dagger = \mathbf{e}^\dagger \mathbf{e} = 0$. The change of coordinates

$$x_1 = \frac{1}{2}(z_1 + z_2)$$

$$x_2 = \frac{1}{2}(z_1 - z_2)$$

allows to write $x = x_1 u_1 + x_2 u_2 = z_1 \mathbf{e} + z_2 \mathbf{e}^\dagger = z$. At this point we can identify the element $z = z_1 \mathbf{e} + z_2 \mathbf{e}^\dagger$ with the pair $(z_1, z_2) \in \mathbb{R}^2$. We note that given two elements $z = z_1 \mathbf{e} + z_2 \mathbf{e}^\dagger$, $z' = z_1' \mathbf{e} + z_2' \mathbf{e}^\dagger$ their sum and product are given by

$$z + z' = (z_1 + z_1')\mathbf{e} + (z_2 + z_2')\mathbf{e}^\dagger, \qquad zz' = (z_1 z_1')\mathbf{e} + (z_2 z_2')\mathbf{e}^\dagger$$

which, at level of pairs, corresponds to the sum and multiplication componentwise. In this new basis, the monogenicity condition (left or right) of a function $w = w(z)$ rewrites as

$$(\mathbf{e}\ \ \mathbf{e}^\dagger) \begin{pmatrix} \dfrac{\partial w_1}{\partial z_1} & \dfrac{\partial w_1}{\partial z_2} \\[2mm] \dfrac{\partial w_2}{\partial z_1} & \dfrac{\partial w_2}{\partial z_2} \end{pmatrix} \begin{pmatrix} \mathbf{e} \\ \mathbf{e}^\dagger \end{pmatrix} = 0,$$

which leads to

$$\frac{\partial w_1}{\partial z_1}\mathbf{e} + \frac{\partial w_2}{\partial z_2}\mathbf{e}^\dagger = 0,$$

that is

$$\frac{\partial w_1}{\partial z_1} = \frac{\partial w_2}{\partial z_2} = 0.$$

Example 3.4 As an example of third order algebra, we consider the ternions. The matrix C is in this case

$$C = \begin{pmatrix} x_1 & 0 & 0 \\ 0 & x_2 & 0 \\ x_3 & 0 & x_2 \end{pmatrix},$$

and $|C| = x_1 x_2^2$. The matrix is lower triangular, with eigenvalues equal to the diagonal elements. Easy arguments show that C can be diagonalized over the real and so the system is completely hyperbolic. We note that by multiplying Ω by its adjoint Ω^* we get that the components y_ℓ, $\ell = 1, 2, 3$ of a monogenic function $y_1 e_1 + y_2 e_2 + y_3 e_3$ satisfy the equation

$$\frac{\partial^3}{\partial x_1 \partial^2 x_2} y_\ell = 0, \qquad \ell = 1, 2, 3.$$

Example 3.5 Let us now consider two cases of four dimensional algebras. First we look at the case of the algebra of quaternions which is a division algebra. Thus we expect an elliptic system. And in fact the left monogenicity condition is associated with the matrix

$$C = \begin{pmatrix} x_1 & -x_2 & -x_3 & -x_4 \\ x_2 & x_1 & -x_4 & x_3 \\ x_3 & x_4 & x_1 & -x_2 \\ x_4 & -x_3 & x_2 & x_1 \end{pmatrix}.$$

We have that $|C| = (x_1^2 + x_2^2 + x_3^2 + x_4^2)^2$ which vanishes only at $(0, 0, 0, 0)$. In the case of the algebra of bicomplex numbers, we have:

$$C = \begin{pmatrix} x_1 & -x_2 & -x_3 & x_4 \\ x_2 & x_1 & -x_4 & -x_3 \\ x_3 & -x_4 & x_1 & -x_2 \\ x_4 & x_3 & x_2 & x_1 \end{pmatrix}$$

and

$$|C| = ((x_1 - x_4)^2 + (x_2 + x_3)^2)((x_1 + x_4)^2 + (x_2 - x_3)^2).$$

The determinant $|C|$ can vanish also for $(x_1, x_2, -x_2, x_1)$ and $(x_1, x_2, x_2, -x_1)$ with $x_1 x_2 \neq 0$. Thus the system is parabolic and it is possible to construct a change of basis for which the determinant of new matrix C associated with the monogenicity condition do not depend on all the four variables.

Remark 3.2 The work of Sce discussed in this chapter was, unfortunately, completely forgotten despite its relations with physical problems, see [3–5]. In the works of Krasnov, see e.g. [6, 7], the author discusses various properties of PDEs in algebras. In particular, in section 7.5 of [6] he considers ellipticity properties of the solutions of a generalized Cauchy–Riemann operator, i.e. monogenic functions, according to the type of algebras considered. In particular, it is shown that an operator is elliptic if and only if the algebra is a division algebra. He also widely discusses the case of the Riccati equation. Among various results, he shows that

Proposition 3.1 (Proposition 2.1, [6]) *Any n-dimensional polynomial differential system $x' = P(x)$ with $\deg P = m$ can be embedded into a Riccati equation considered in an algebra $\mathscr{A}$ of dimension $\geq n$.*

References

1. Albert, A.A.: Structure of Algebras, Vol. XXIV. American Mathematical Society Colloqium Publications, New York (1939)
2. Albert, A.A.: Modern Higher Algebra. The University of Chicago Science Series, Chicago (1937)
3. Courant, R., Friedrichs, K.O.: Supersonic Flow and Shock Waves. Interscience Publishers, New York, NY (1948)
4. Courant, R., Hilbert, D.: Methoden der Mathematischen Physik. Band II, Berlin (1937)
5. Courant, R., Lax, P.: On nonlinear partial differential equations with two independent variables. Comm. Pure Appl. Math. **2**, 255–273 (1949)
6. Krasnov, Y.: Differential equations in algebras. Hypercomplex Analysis, pp. 187–205. Trends Math. Birkhäuser Verlag, Basel (2009)
7. Krasnov, Y.: Properties of ODEs and PDEs in algebras. Complex Anal. Oper. Theory **7**, 623–634 (2013)
8. Sce, M.: Monogeneità e totale derivabilità nelle agebre reali e complesse. Atti Accad. Naz. Lincei. Rend. Cl. Sci. Fis. Mat. Nat., (8) **16**, 30–35 (1954). See also this volume, Chap. 2
9. Sce, M.: Su alcune proprietà delle matrici permutabili e diagonalizzabili. Rivista di Parma **1**, 363–374(1950)
10. Scorza, G.: Corpi numerici e algebre. Messina (1921)
11. Scorza, G.: La teoria delle algebre e sue applicazioni. Atti 1° Congresso dell'Un. Mat. Ital., 40–57 (1937) (available on line)
12. Segre, B.: Forme differenziali e loro integrali. Roma (1951)
13. Wedderburn, J.H.M.: A type of primitive algebra. Trans. Am. Math. Soc., **15**, 162–166 (1914)

Chapter 4
On the Variety of Zero Divisors in Algebras

This chapter contains the translation of the paper:

M. Sce, *Sulla varietà dei divisori dello zero nelle algebre*, (Italian) Atti Accad. Naz. Lincei Rend. Cl. Sci. Fis. Mat. Nat. (8) 23 (1957), 39–44.

Article by Michele Sce, presented during the meeting of 8 August 1957 by B. Segre, member of the Academy.

In this work, after some brief consideration on matrices on (skew) fields, we show how the study of the variety of zero divisors in alternative algebras on fields of characteristic different from 2 and in Jordan algebras on characteristic zero fields, can be led to the study of these varieties in simple, central algebras.

We then show that *the dimension of the variety of zero divisors in alternative algebras on fields of characteristic different from 2 is given by the order of the algebra minus the order of the smallest primitive algebra factor of the simple components of the algebra.*

Finally, based on the known classification, we give some partial result on Jordan algebras.

1. Let A be a matrix of order n on a field **C** whose element in the first row and first column is nonzero [Editors' Note: elsewhere **C** may denote the complex field, here it denotes *any* field]; then

$$A = \begin{pmatrix} a_{11} & a_{12} & \cdots & a_{1n} \\ a_{21} & a_{22} & \cdots & a_{2n} \\ \cdots & \cdots & \cdots & \cdots \\ a_{n1} & a_{n2} & \cdots & a_{nn} \end{pmatrix} =$$

F. Colombo et al., *Michele Sce's Works in Hypercomplex Analysis*, https://doi.org/10.1007/978-3-030-50216-4_4

$$= \begin{pmatrix} a_{11} & 0 & \cdots & 0 \\ a_{21} & a_{22} - a_{21}a_{11}^{-1}a_{12} & \cdots & a_{2n} - a_{21}a_{11}^{-1}a_{1n} \\ \cdots & \cdots & \cdots & \cdots \\ a_{n1} & a_{n2} - a_{n1}a_{11}^{-1}a_{12} & \cdots & a_{nn} - a_{n1}a_{11}^{-1}a_{1n} \end{pmatrix} \begin{pmatrix} 1 & a_{11}^{-1}a_{12} & \cdots & a_{11}^{-1}a_{1n} \\ 0 & 1 & \cdots & 0 \\ \cdots & \cdots & \cdots & \cdots \\ 0 & 0 & \cdots & 1 \end{pmatrix} = BH.$$

If A has all the elements in the first column different from zero, we can bring to the first place any row and decompose the new matrix in the same way we decomposed A; in this way, starting from A, we obtain n matrices having all the elements in the first row nonzero, except for the first one. Continuing in this way, we can write A— under suitable convenient qualitative hypothesis on the elements of the first $n - 1$ columns—in triangular form, in $n!$ ways.[1]

If we do not make any assumption on the elements of A, some or none of the expressions giving A may have sense; in the first case we will call values of A the elements of $\mathbf{C}$ given by the expressions having sense, in the second case we will say that A has value zero. Sometimes, in the sequel, we will write $|A| = 0$ to indicate that A has value zero.[2]

Using the relation that we established between the values of A and the reduction of A to triangular form, one may easily show that if the columns of A are linearly dependent on the right, all the values of A are zero and that—viceversa—if a value of A is zero, the columns of A are linearly dependent on the right (and thus also the other values of A are zero). Then, by using known theorems on linear systems,[3] one can show that *the values of the matrix are zero if and only if its rows are linearly dependent* (on the left or on the right).

From this it follows that *the values of the product of two matrices are zero if and only if the values of one of the two matrices are zero.* It is however easy to verify that the values of the product are not, in general, the product of the values of the factor matrices.

2. Let $\mathbf{A}$ be an algebra of order nd, with unit, on a field $\mathbf{F}$, which contains a primitive, associative algebra $\mathbf{C}$ (which may coincide with $\mathbf{F}$) of order d and an algebra $\mathbf{B}$, with unit, and with basis $(u_1, \ldots, u_n)$. If the elements of $\mathbf{C}$ can be associated with the elements in $\mathbf{A}$ and can be commuted with the elements in $\mathbf{B}$,

---

[1] With the same procedure we can write symmetric matrices in the canonical form

$$\begin{pmatrix} D & 0 \\ 0 & E \end{pmatrix}, \qquad \text{with } E = \begin{pmatrix} 0 & 1 \\ 1 & 0 \end{pmatrix} \times D',$$

where $\times$ denotes the direct product of matrices and D, D' are diagonal.

[2] If the field is commutative, all the values of the matrix coincide with the classical determinant (of course, the values may coincide among them even if the field is not commutative). Starting from considerations similar to ours, J. Dieudonné, [*Les déterminants sur un corps non commutatif*, Bull. Soc. Math. France, **74** (1943), pp. 27–45], introduces a determinant sharing most of the classical properties, but not in the (skew) field. These considerations extend immediately to fields with associative inverse (alternative fields).

[3] See B. Segre, *Lezioni di Geometria moderna*, vol. I, (Bologna, 1947), n. 97.

i.e., if $\mathbf{C}$ is in the kernel of $\mathbf{A}$ and in the commutator of $\mathbf{B}$, we will say that $\mathbf{A}$ is direct product of $\mathbf{B}$ and $\mathbf{C}$, and we will write $\mathbf{A} = \mathbf{B} \times \mathbf{C}$. We will write the elements in $\mathbf{A}$ as

$$x = x_1 u_1 + \cdots + x_n u_n = \xi \mathbf{u}_{-1} = \mathbf{u}\xi_{-1}$$

$$y = y_1 u_1 + \cdots + y_n u_n = \mu \mathbf{u}_{-1} = \mathbf{u}\mu_{-1} \qquad (x_i, y_i \in \mathbf{C}).$$

If x is a left or right zero divisor, there exists y such that

$$0 = xy = x\mathbf{u}\eta_{-1} = \mathbf{u}X'\eta_{-1} \iff X'\eta_{-1} = 0 \qquad (4.1)$$

or

$$0 = yx = \xi \mathbf{u}_{-1}x = \xi X''\mathbf{u}_{-1} \iff \xi X'' = 0 \qquad (4.2)$$

where X' (X'') represents x in the $\mathbf{C}$-module $\mathbf{A}'$ ($\mathbf{A}''$) first and second representation of $\mathbf{A}$. Relations (4.1), (4.2) say that *x is a left or right zero divisor if and only if the values of X' or of X'' are zero* and we will say that these values are first and second values of x over $\mathbf{C}$; the values of x over $\mathbf{F}$ are called first and second norm $n'(x)$, $n''(x)$ of x.

If $\mathbf{C}$ is, in turn, the direct product of a skew field $\mathbf{C}_1$ and a field $\mathbf{C}_2$, the elements of $\mathbf{A}'$ and $\mathbf{A}''$ consist of matrices which are the first and second representation of $\mathbf{C}_1$ over $\mathbf{C}_2$. Moreover, this ensures that the norms of x are the values over $\mathbf{F}$ of the values of x over $\mathbf{C}$.

Let now $\mathbf{R}$ be an ideal of $\mathbf{A}$ and let

$$\mathbf{S} = (u_{m+1}, \ldots, u_n) = \mathbf{u}^{(2)}$$

the ideal of

$$\mathbf{B} = (\mathbf{u}^{(1)} \ \mathbf{u}^{(2)})$$

trace of $\mathbf{R}$ on $\mathbf{B}$. Let us consider the first and second representation of an element in $\mathbf{A}$, $x = x^{(1)} + x^{(2)}$ ($x^{(2)} \in \mathbf{R}$) over $\mathbf{C}$:

$$x\mathbf{u} = (x^{(1)} + x^{(2)})(\mathbf{u}^{(1)} \ \mathbf{u}^{(2)}) = (\mathbf{u}^{(1)} \ \mathbf{u}^{(2)}) \begin{pmatrix} X_1^{(1)} & 0 \\ X_2^{(1)} + X_1^{(2)} & X_1^{(1)} + X_2^{(2)} \end{pmatrix} = \mathbf{u}X'$$

[Editors' Note: the entry (2, 2) in the matrix is $X_3^{(1)} + X_2^{(2)}$ in the original manuscript] where $X_i^{(k)}$ depends on the coordinates of $x^{(k)}$ ($k = 1, 2$); $X_1^{(1)}$ represents the class $[x^{(1)}]$ in $\mathbf{A} - \mathbf{R}$.

If $x^{(1)}$ is a zero divisor in $\mathbf{A} - \mathbf{R}$, $X_1^{(1)}$ has dependent columns and thus dependent rows; then $|X'| = 0$ and x is a left zero divisor in $\mathbf{A}$ for any $x^{(2)}$ in $\mathbf{R}$.

Therefore, *the variety of left zero divisors in* **A** *has always at least dimension*

$$v' + (n - m)d \tag{4.3}$$

where $(n - m)d$ *is the dimension of the space representing the ideal* **R** *and* v' *is the dimension of the variety* $V_0'(\mathbf{A} - \mathbf{R})$ *of the left zero divisors of* $\mathbf{A} - \mathbf{R}$. [4]

If $\mathbf{A} = \mathbf{A}_1 \oplus \mathbf{A}_2$ is direct sum of two algebras $\mathbf{A}_1, \mathbf{A}_2$ of order n_1 and n_2, taking into account that $\mathbf{A}_1$ and $\mathbf{A}_2$ are ideals of $\mathbf{A}$ and that the vector spaces $\mathbf{A}, \mathbf{A}', \mathbf{A}''$ are isomorphic, we conclude that *the dimension of* $V_0'(\mathbf{A})$ *is the greatest of the two numbers* $n_1 + \dim V_0(\mathbf{A}_2)$ *and* $n_2 + \dim V_0(\mathbf{A}_1)$.

3. Let $\mathbf{A}$ be a t-algebra, namely an algebra which admits as trace function the trace of the elements in the $\mathbf{F}$-module of the representation. Then we can define the radical (maximal nilpotent ideal) of $\mathbf{A}$ as the set of elements h such that the trace $t(xh)$ vanishes for every x in $\mathbf{A}$. It follows that $t(x^i h^k) = 0$ for every pair of positive integers i and k; thus $t[(x + h)^j]$ as well as the norm $n(x + h)$ do not depend on the coordinates of h.

We can then conclude that *the zero divisors of a t-algebra* $\mathbf{A}$ *which are not in the radical* $\mathbf{R}$ *come from zero divisors of the semisimple part* $\mathbf{A} - \mathbf{R}$; *thus the dimension of* $V_0(\mathbf{A})$ *is exactly the one given in* (4.3).

Taking into account that semisimple algebras are direct sums of simple algebras, one also has that $\dim V_0'(\mathbf{A})$ *is completely determined when the dimension of the variety of zero divisors of simple t-algebras is known.*

Since it is known that the radical of alternative algebras over fields of characteristic different from 2 consists of elements h such that $x + h$ is a zero divisor if and only if x is a zero divisor,[5] for these algebras the theorems stated above for t-algebras are valid.

In force of these theorems, we can reduce ourselves to the study of the varieties of zero divisors in simple algebras; if then—as we will do in the sequel—we will consider only algebras whose center is a field, we can reduce ourselves to the consideration of central simple algebras.

4. Let $\mathbf{A}$ be a central, simple, associative algebra over a field $\mathbf{F}$; then $\mathbf{A} = \mathbf{M} \times \mathbf{C}$, where $\mathbf{M}$ is a regular algebra of degree r and $\mathbf{C}$ is a primitive algebra of order d. In order for an element of $\mathbf{A}$ to be a zero divisor it is necessary and sufficient that its values are zero in $\mathbf{C}$; thus

$$\dim V_0(\mathbf{A}) \geq r^2 d - d. \quad [6] \tag{4.4}$$

---

[4] When $\mathbf{A} - \mathbf{R}$ is a subalgebra of $\mathbf{A}$, these considerations have a simple geometric interpretation: $V_0'(\mathbf{A})$ contains a cone- of dimension $v' + (n - m)d$ - having $V_0'(\mathbf{A} - \mathbf{R})$ as vertex.

[5] See R. Dubisch, S. Perlis, *The radical of an alternative algebra*, Amer. J. Math., **70** (1948), pp. 540–546.

[6] In the case of associative and alternative algebras it is not necessary to distinguish left and right zero divisors. One can arrive to (4.4) observing that, given the shape of the matrices in $\mathbf{A}'$ (or in $\mathbf{A}''$), $V_0(\mathbf{A})$ contains for sure two sets depending on $(r-1)d$ parameters of $(r^2 d - rd)$-dimensional linear spaces.

Alternative, simple, central algebras which are not associative are Cayley-Dickson algebras; an element $x = \sum_{i=1}^{8} x_i u_i$ ($x_i \in \mathbf{F}$) of such algebras is a zero divisor if and only if it is zero the expression

$$x_1^2 - (\alpha x_2^2 + \beta x_3^2 - \alpha\beta x_4^2) - \gamma[x_5^2 - (\alpha x_6^2 + \beta x_7^2 - \alpha\beta x_8^2)], \quad (\alpha, \beta, \gamma \in \mathbf{F}). \quad (4.5)$$

When $x' = \sum_{i=1}^{4} x_i u_i$, $x'' = \sum_{i=5}^{8} x_i u_i$ are elements of a primitive algebra, (4.5) vanishes if and only if

$$\gamma = z_1^2 - \alpha z_2^2 - \beta z_3^2 + \alpha\beta z_4^2 \quad (z_i \in \mathbf{F}).^7 \quad (4.6)$$

(Editors' note: $\alpha\beta z_4^2$ was $\alpha\beta_4^2$ in the original manuscript). If (4.6) cannot be satisfied, the algebra is primitive. On the contrary, if (4.6) admits solutions, the quadric cone of the 8-dimensional space obtained by equating (4.5) to zero contains a linear, 4-dimensional space (on $\mathbf{F}$) and is a hypersurface (on $\mathbf{F}$).

In the case x' and x'' do not belong to a primitive algebra, it will happen e.g. that

$$\beta = t_1^2 - \alpha t_2^2 \quad (t_i \in \mathbf{F});$$

but we find again that the cone $V_0(\mathbf{A})$ contains a 4-dimensional space and is a hypersurface.

Now taking into account that—by the theorem of Dubisch-Perlis—if $\mathbf{F}$ has characteristic different from 2, the values of elements of $\mathbf{A}$ depend on all the coordinates, we have that *if $\mathbf{A}$ is an (associative or) alternative simple algebra on a field with characteristic different from 2*

$$\dim V_0(\mathbf{A}) = n - d$$

where n is the order of $\mathbf{A}$ and d is the order of the primitive algebra with maximum order which is factor of $\mathbf{A}$.

From here—using the theorems in n. **2** and **3**—it immediately follows the theorem stated in the introduction.

5. Central, simple, Jordan algebras over a field $\mathbf{F}$ of characteristic zero are of the following four types:

a) Algebras of degree 2 with basis $u_1, \ldots, u_n$ ($n \geq 4$) and the multiplication table

$$u_1 u_i = u_i u_1 = u_i, \quad u_i^2 = \alpha_i u_1 \quad (\alpha_i \text{ in } \mathbf{F}), \quad u_i u_j = u_j u_i = 0.$$

---

[7] For all these notions, see L. E. Dickson, *Algebren und ihre Zahlentheorie*, Zurich 1927, Kap. XII, § 133. Note that the norm of elements of a Cayley-Dickson algebra, defined in n. **2**, is the fourth power of (4.5).

b) Algebras $\mathbf{A}^+$ which can be obtained from a central, simple, associative algebra $\mathbf{A} = \mathbf{M} \times \mathbf{C}$, ($\mathbf{M}$ regular algebra of degree r, $\mathbf{C}$ primitive algebra of order d) considering as new multiplication the operation

$$x \cdot y = \frac{xy + yx}{2}. \tag{4.7}$$

c) Algebras $\mathbf{S}^+$ having—as elements—the symmetric elements of a central, simple, associative algebra $\mathbf{A} = \mathbf{M} \times \mathbf{C}$ having an involution J and—as multiplication— (4.7).

d) Algebras of order 27 having as elements hermitian matrices of the third order over Cayley algebras and multiplication (4.7).

Jordan matrices of type a) have, evidently, n hyperplanes of zero divisors.

When one considers the representation of Jordan algebras of type b) over $\mathbf{F}$ one obtains

$$2x\mathbf{u}_{-1} = x\mathbf{u}_{-1} + \mathbf{u}_{-1}x = (X'_{-1} + X'')\mathbf{u}_{-1}; \tag{4.8}$$

x is a zero divisor if and only if its norm vanishes, i.e. $|X'_{-1} + X''| = 0$. The matrix $X'_{-1} + X''$ is formed with the matrices representing elements in the associative algebra $\mathbf{C}$ and in the Jordan algebra $\mathbf{C}^+$. If $\mathbf{C}^+$ is a primitive algebra, the norm of x is the norm of an element in $\mathbf{C}^+$ and thus its vanishing implies at most d conditions. An easy check shows that the conditions are exactly d.

The situation is less simple if $\mathbf{C}^+$ is not primitive; then $\mathbf{C}^+$ possesses already a variety of zero divisors and it turns out that:[8]

$$V_0(\mathbf{C}^+) = d - \frac{s}{2} \tag{4.9}$$

where s is the degree of $\mathbf{C}$ (degree of the minimum equation of $\mathbf{C}$).

To approach the solution of the problem it seems to us necessary to look for a canonical form for matrices on any skew field.[9]

A first result—in this stream of ideas—is that *the zero divisors of $\mathbf{A}$ are the zero divisors of $\mathbf{A}^+$*,

$$V_0(\mathbf{A}^+) \supset V_0(\mathbf{A});$$

then $V_0(\mathbf{A}^+)$ contains a variety $W(\mathbf{A})$ made by all zero divisors of $\mathbf{A}^+$ which are not zero divisors of $\mathbf{A}$. Our problem reduces to the problem of finding the dimension of $W(\mathbf{A})$.

---

[8] See C. M. Price, *Jordan division algebras and the algebras $A(\lambda)$*, Trans. Amer. Math. Soc., **70** (1951), pp. 291–300.

[9] For the preliminary notions see, for example, See J. L. Brenner, *Matrices of quaternions*, Pacific J. Math., **1** (1951), pp. 329–335, n. 2 and 3.

When one takes in **M** a basis of symmetric elements, we can represent through (4.8) also elements of the algebras of type c) and d). For algebras of type c) we have to repeat the arguments for algebras of type b). On the other hand, for algebras of type d), since the symmetric elements of the Cayley algebra form now the field **F**, we can say that $V_0(\mathbf{A})$ is a hypersurface.

Taking into account that—provided (4.9)—a Jordan algebra coming from a primitive algebra is not necessarily primitive if its degree, and so its order, is even we can say that *for simple Jordan algebras of type a) and d) and for those of odd order and type b and c) the equality*

$$\dim V_0(\mathbf{A}) = n - d$$

holds, where n is the order of the algebra **A** *and d is the order of the primitive algebra of maximum order contained in* **A**.

From this one could obtain results for non simple Jordan algebras, but we do not insist on these.

4.1 Comments and Historical Remarks

In this paper, Sce considers algebras which are alternative over fields with characteristic different from 2 and in Jordan algebras over fields of characteristic 0. This study looks very pioneering at his time, and to the best of our knowledge, there are few works in the literature treating the description of zero divisors in Clifford algebras. The study of zero divisors is of crucial importance while dealing with notions like the one of derivability and the Cauchy integral formula.

We begin by recalling the notion of Jordan algebra which was begun by Jordan, von Neumann and Wigner in order to formulate the foundations of quantum mechanics in terms of a suitable product, instead of the usual one. Other useful references for the sequel are [1, 6–9, 11, 17].

Definition 4.1 A Jordan algebra is a nonassociative algebra $\mathscr{A}$ over a field F of characteristic different from 2 whose multiplication satisfies:

1. $a \cdot b = b \cdot a, \forall a, b \in \mathscr{A}$;
2. $(a \cdot b) \cdot (a \cdot a) = a \cdot (b \cdot (a \cdot a)), \forall a, b \in \mathscr{A}$, (Jordan law).

Remark 4.1 We point out that a Jordan algebra is power associative, namely the subalgebra generated by any element is associative. Thus, when considering products of an element with itself, it does not matter how the operations are carried out so, for example, $x \cdot ((x \cdot x) \cdot x) = (x \cdot x) \cdot (x \cdot x) = x \cdot (x \cdot (x \cdot x))$ etc.

Moreover, given an associative algebra $\mathscr{A}$ one can construct an algebra, denoted by $\mathscr{A}^+$, having the same underlying vector space as $\mathscr{A}$ and the multiplication given by

$$a \circ b = \frac{a \cdot b + b \cdot a}{2},$$

where $a\dot{b}$ denotes the multiplication in $\mathscr{A}$. The multiplication $a \circ b$ is called Jordan product. Obviously, an associative algebra is a Jordan algebra if and only if it is commutative.

A Jordan algebra is said to be a *special Jordan algebra* if it is an algebra of the form $\mathscr{A}^+$ or one of its subalgebras. Otherwise it is called an *exceptional Jordan algebra*. As examples of special Jordan algebras we can take the set of self-adjoint real, complex, or quaternionic matrices with the Jordan multiplication. As example of exceptional Jordan algebra, is the set of 3×3 self-adjoint matrices over the octonions, with the Jordan multiplication. It has dimension 27 and it is an example of algebras of order 27 considered in the previous section, n. **5**.

We also recall the following notion which also appears in the previous section:

Definition 4.2 An algebra over a field F, with unit u_1 is called *central* if its center, namely the set of elements commuting with any other element in the algebra, coincides with $Fu_1 = \{ku_1 \mid k \in F\}$.

The study of zero divisors performed by Sce is refined and done in a rather general setting. As we said already, in the framework of Clifford algebras a lot is known about the analysis of functions, in particular monogenic functions, but there is no systematic study of zero divisors. Some studies have been done for bicomplex and biquaternionic numbers, for hyperbolic numbers, see [10, 14, 16], in the case of the Clifford algebra $\mathbb{R}_3$, see [13]. Let us recall the following well known fact:

Proposition 4.1 *Let $\mathbb{R}_n(= \mathbb{R}_{0,n})$ be the Clifford algebra generated by $e_1, \ldots, e_n$ satisfying $e_i e_j + e_j e_i = -2\delta_{ij}$, $i, j = 1, \ldots n$, where δ_{ij} is the Kronecker's delta. For $n \geq 3$ the Clifford algebra contains zero divisors.*
Let $\mathbb{R}_{p,q}$, $p + q = n$, be the Clifford algebra generated by $e_1, \ldots, e_n$ satisfying $e_i^2 = 1$, $i = 1, \ldots, p$, $e_i^2 = -1$, $i = p + 1, \ldots, n$, $e_i e_j + e_j e_i = 0$, $i \neq j$, $i, j = 1, \ldots, n$. For $p > 0$ the Clifford algebra $\mathbb{R}_{p,q}$ contains zero divisors.

Example 4.1 In the case of bicomplex numbers, let us consider the basis $\{1, i, j, ij\}$ where $i^2 = j^2 = -1$, $ij = ji$. The bicomplex numbers may be considered complex numbers with complex coefficients. We can write them in various ways:

$$Z = z_1 + jz_2, \qquad z_1, z_2 \in \mathbb{C}_i$$
$$Z = w_1 + iw_2, \qquad w_1, w_2 \in \mathbb{C}_j$$

where $\mathbb{C}_i$, $\mathbb{C}_j$ are the complex planes with imaginary units i, j, respectively. Let us introduce the notation

$$|Z|_i^2 = z_1^2 + z_2^2$$

Note that $|\cdot|_i$ is a $\mathbb{C}_i$-valued function. There are three possible conjugations of the bicomplex number Z one of which is defined by

$$Z^\dagger = z_1 - jz_2.$$

With this definition it is clear that

$$|Z|_i^2 = ZZ^\dagger \tag{4.10}$$

thus if $Z \neq 0$ but $|Z|_i = 0$ then it is clear that Z is a zero divisor since also $Z^\dagger$ is nonzero. If $|Z|_i \neq 0$ then Z is invertible and (4.10) shows that

$$Z^{-1} = \frac{Z^\dagger}{|Z|_i^2}.$$

The variety of zero divisors is expressed by the equation $z_1^2 + z_2^2 = 0$ which is equivalent to $z_1 = \pm i z_2$. We conclude that all the zero divisors in the algebra of bicomplex numbers are of the form

$$Z = \lambda(1 \pm ij), \qquad \lambda \in \mathbb{C}_i \setminus \{0\}.$$

There are equivalent ways to express the zero divisor, and we refer the reader to [10] for more information. One that is very useful is the one based on the so-called idempotent decomposition. Let us consider

$$\mathbf{e} = \frac{1 + ij}{2}, \qquad \mathbf{e}^\dagger = \frac{1 - ij}{2},$$

that are such that $\mathbf{e}^2 = \mathbf{e}$, $\mathbf{e}^{\dagger^2} = \mathbf{e}^\dagger$, $\mathbf{e}\mathbf{e}^\dagger = 0$ and $\mathbf{e} + \mathbf{e}^\dagger = 1$. Then, every bicomplex number can be written as

$$Z = \beta_1 \mathbf{e} + \beta_2 \mathbf{e}^\dagger,$$

with $\beta_1, \beta_2 \in \mathbb{C}_i$ (or in $\mathbb{C}_j$) and the zero divisors are all of the form $\beta_1 \mathbf{e}$, $\beta_2 \mathbf{e}^\dagger$, $\beta_1, \beta_2 \neq 0$.

Example 4.2 With the notations of the previous example, let us set $k = ij$ and consider

$$\mathbb{D} = \{\zeta = x + ky \mid x, y \in \mathbb{R}\}.$$

This is the set of hyperbolic numbers. Let us introduce the notation

$$|\zeta|_h^2 = x^2 - y^2.$$

We note that

$$|\zeta|_h^2 = (x + ky)(x - ky) = \zeta\zeta^\circ$$

where we set $z^{\circ} = x - ky$. The value of $|\cdot|_h^2$ is a real number which can be negative and can vanish for $\zeta \neq 0$, more precisely it vanishes for $x = \pm y$. In this case, when $\zeta \neq 0$ and $\zeta \zeta^{\dagger} = 0$ then ζ is a zero divisor and this happens if and only if $\zeta = \lambda(1 \pm k)$, $\lambda \in \mathbb{R} \setminus \{0\}$.

Example 4.3 The only case in which the zero divisors in a Clifford algebra have been fully described is the case of the Clifford algebra $\mathbb{R}_3$ over three imaginary units e_1, e_2, e_3 satisfying $e_i^2 = -1$, $e_i e_j + e_j e_i = 0$, for all $i, j = 1, 2, 3, i \neq j$. This case is studied by Rizza in [13]. He first compute the product of two elements $x, y \in \mathbb{R}_3$ by writing $x = q + q'e_3$, where q, q' are in the algebra generated by e_1, e_2 which is isomorphic to the algebra of quaternions, and then rewriting

$$q = \alpha_1 + \beta_1 e_2, \qquad q' = \alpha_1' + \beta_1' e_2,$$

where $\alpha_1, \beta_1, \alpha_1', \beta_1' \in \mathbb{C}_{e_1}$, the complex plane with imaginary unit e_1. He shows that x is a zero divisor if and only if $\alpha_1 = \pm e_1 \beta_1'$, $\beta_1 = \pm e_1 \alpha_1'$. Thus the zero divisors form a linear variety of real dimension 4.

In alternative, one can introduce the two orthogonal idempotents

$$\mathsf{o} = \frac{1 + e_{123}}{2}, \qquad \mathsf{o}^{\dagger} = \frac{1 - e_{123}}{2}$$

and observe that x is a zero divisor if and only if it is of the form $x = q\mathsf{o}$, or $x = q\mathsf{o}^{\dagger}$, where q is a nonzero quaternion.

Remark 4.2 A full characterization of general Clifford algebras $\mathbb{R}_{p,q}$ is available and it shows that they are either algebras of matrices over $\mathbb{R}$, $\mathbb{C}$, $\mathbb{H}$ or direct sum of such matrices, see [6]. By Bott periodicity, it is enough to describe the case $p + q < 8$. In view of this result, any element of a Clifford algebra can be identified with a matrix with real entries and so it is invertible, i.e. it is not a zero divisor, if and only if its determinant is nonzero.

The knowledge of zero divisors is crucial when dealing with two notions in analysis: the one of derivability, which require the construction of the limit of a suitable difference quotient, and integral formulas like the Cauchy integral formula.

As we discussed in Chap. 2, in the hypercomplex setting one can define a generalization of holomorphic functions of a complex variables in various ways: one is to consider functions which can be expanded in converging power series of the variable, the second is to consider functions that admit in each point derivative in all directions and the third is to consider functions in the kernel of a suitable operator generalizing the Cauchy-Riemann operator. In all three cases there will be a theory on the left and one on the right, if the algebra is noncommutative. In the second and third case the zero divisors play a role and this was also a reason to study them, see [12].

When considering the notion of derivative in a given direction, one has to deal with difference quotients left or right, namely with quotients of the form

$$(\Delta x)^{-1}(\Delta f) \quad \text{or} \quad (\Delta f)(\Delta x)^{-1}$$

and in order to have well defined quotients, it is necessary that Δx is invertible and, in particular, it is not a zero divisor (we always assume that the algebra has a unit). In the case of integral formulas one has to pay attention to the fact that there may exists contour on integration which are homologically trivial in an open set of the hypercomplex algebra, but it is not necessarily true that this contour is homologically trivial when considered in the same open set minus the zero divisors. Thus it is necessary to add topological hypothesis on the open set. We refer the reader to [3–5, 12] for more information.

Remark 4.3 As it is well known Clifford analysis has been widely studied since the seventies, when the Belgian school around Brackx, Delanghe and Sommen and later also the Czech school around Bures and Souček started the systematic study of functions in the kernel of suitable operators, generalizing the Cauchy-Riemann operator to higher dimensions, see [2, 6] and the references therein. All these fruitful studies were performed considering functions defined on open sets in $\mathbb{R}^n$ (or $\mathbb{R}^{n+1}$) with values in the Clifford algebra $\mathbb{R}_n$ over n imaginary units. The elements in $\mathbb{R}^n$ are identified with the so-called 1-vectors in the algebra, i.e. $(x_1, \ldots, x_n) \mapsto x_1 e_1 + \cdots + x_n e_n$ while elements in $\mathbb{R}^{n+1}$ are identified with the so-called paravectors in the algebra, i.e. $(x_0, x_1, \ldots, x_n) \mapsto x_0 + x_1 e_1 + \cdots + x_n e_n$. In this way, one completely avoids the problems of having zero divisors as input of the functions. Clifford analysis has been generalized to the complexified case by the above authors and also by Ryan, see [15] but again considering functions $\mathbb{C}^n$ or $\mathbb{C}^{n+1}$ with values in the Clifford algebra generated by n imaginary units, over the complex field.

References

1. Brenner, J.L.: Matrices of quaternions. Pac. J. Math., **1**, 329–335 (1951)
2. Brackx, F., Delanghe, R., Sommen, F.: Clifford analysis. In: Research Notes in Mathematics, vol. 76. Pitman (Advanced Publishing Program), Boston, MA (1982)
3. Bures, J., Souček, V.: Generalized hypercomplex analysis and its integral formulas. Complex Variables **5**, 53–70 (1985)
4. Colombo, F., Loustaunau, P., Sabadini, I., Struppa, D.C.: Regular functions of biquaternionic variables and Maxwell's Equations. J. Geom. Phys. **26**, 183–201 (1998)
5. Colombo, F., Sabadini, I., Sommen, F., Struppa, D.C.: Analysis of Dirac systems and computational algebra. In: Progress Mathematical Physics, vol. 39. Birkhäuser, Boston (2004)
6. Delanghe, R., Sommen, F., Souček, V.: Clifford algebra and Spinor-valued functions. In: Math. and Its Appl., vol. 53. Kluwer Acad. Publ., Dordrecht (1992)
7. Dubisch, R., Perlis, S.: The radical of an alternative algebra. Am. J. Math. **70**, 540–546 (1948)
8. Dickson, L.E.: Algebren und ihre Zahlentheorie. Zurich (1927)
9. Dieudonné, J.: Les déterminants sur un corps non commutatif. Bull. Soc. Math. France **74**, 27–45 (1943)

10. Luna–Elizarraras, M.E., Shapiro M., Struppa, D.C., Vajiac, A.: Bicomplex holomorphic functions: the algebra, geometry and analysis of bicomplex numbers. In: Frontiers in Mathematics. Birkhäuser, Springer Cham (2015)
11. Price, C.M.: Jordan division algebras and the algebras $A(\lambda)$. Trans. Am. Math. Soc. **70**, 291–300 (1951)
12. Rizza, G.B.: Sulle funzioni analitiche nelle algebre ipercomplesse. Pont. Acad. Sci. Comment. **14**, 169–194 (1950)
13. Rizza, G.B.: Sulla struttura delle algebre di Clifford. Rend. Seminario Mat. Univ. Padova **23**, 91–99 (1954)
14. Rochon, D., Shapiro, M.: On algebraic properties of bicomplex and hyperbolic numbers. An. Univ. Oradea Fasc. Mat. **11**, 71–110 (2004)
15. Ryan, J.: Complexified Clifford analysis. Complex Variables Theory Appl. **1**, 119–149 (1982/83)
16. Sangwine, S.J., Alfsmann, D.: Determination of the biquaternion divisors of zero, including the idempotents and nilpotents. Adv. Appl. Clifford Algebr. **20**, 401–410 (2010)
17. Segre, B.: Lezioni di Geometria moderna, vol. I. Bologna (1947)

Chapter 5
Remarks on the Power Series in Quadratic Modules

This Chapter contains the translation of the paper:

M. Sce, *Osservazioni sulle serie di potenze nei moduli quadratici*, Atti Accad. Naz. Lincei. Rend. Cl. Sci. Fis. Mat. Nat., (8) **23** (1957), 220–225.

Article by Michele Sce, presented during the meeting of 9 November 1957 by B. Segre, member of the Academy.

In this short paper we consider modules with units which are quadratic, that is, whose elements (with respect to the multiplicative structure induced in the module by their tensor algebra) satisfy a quadratic equation. We show that, in these modules, power series (positive o negative)—if the order of the module is even—are nullsolutions of a power of a generalized laplacian. This fact allows to generalize some results on quaternionic functions of Fueter and his school to Clifford algebras.

1. Let **M** be a module on a field **F** with characteristic not equal 2 and let $1 = i_0, i_1, \ldots, i_n$ be a basis. After identifying the unit of **F** with the unit of **M**, we can write the elements in **M** in the form

$$x = x_0 + x_1 i_1 + \cdots + x_n i_n = x_0 + \mathbf{x} \qquad (x_i \in F).$$

Let **T** be the tensor algebra over **M** and let us assume that for the elements x^2 in **T** one has

$$\mathbf{x}^2 = q(\mathbf{x}) = \sum_{j,k=1}^{n} a_{jk} x_j x_k \qquad (5.1)$$

where $q(\mathbf{x})$ denotes a quadratic form on **F**; it follows that $x^2(\in \mathbf{T})$ is in **M**. Thus **M** is closed with respect to the operation that to the pair x, y associates $\dfrac{xy + yx}{2}$, which gives a Jordan algebra $\mathbf{M}^+$. When one considers the module **M** in **T** equipped

F. Colombo et al., *Michele Sce's Works in Hypercomplex Analysis*, https://doi.org/10.1007/978-3-030-50216-4_5

with the multiplicative structure of $\mathbf{M}^+$, one will say that $\mathbf{M}$ is a quadratic module and will denote it by $\mathbf{M}_q$.

Since, by reducing $q(\mathbf{x})$ to a canonical form, one notices that $\mathbf{M}^+$ is a Jordan algebra, central, simple, of degree 2, then $\mathbf{M}_q$ can be embedded only in algebras $\mathbf{A}$ such that $\mathbf{A}^+$ contains such a Jordan algebra. Among these algebras, those which may be obtained with the Cayley–Dickson process are particularly interesting; these algebras are themselves quadratic modules.[1] If, in addition, $\mathbf{A} \supset \mathbf{M}_q$ is associative, it contains the algebra quotient of $\mathbf{T}$ and of the ideal generated by (5.1); thus the smallest associative algebra containing a quadratic module is a Clifford algebra or an algebra whose semisimple part is a Clifford algebra and whose radical is an algebra with vanishing square—according to the fact that $q(\mathbf{x})$ is degenerate or not.[2]

2. We shall call conjugate of an element $x = x_0 + \mathbf{x}$ in $\mathbf{M}_q$ the element $\bar{x} = x_0 - \mathbf{x}$; it is immediate that

$$x + \bar{x} = 2x_0 = t(x) \qquad \text{(trace of } x)$$

$$x\bar{x} = x_0^2 - q(\mathbf{x}) = n(x) \qquad \text{(norm of } x)$$

are in $\mathbf{F}$ and that the elements x in $\mathbf{M}_q$ satisfy the equation in $\mathbf{F}$

$$z^2 - t(x)z + n(x) = 0. \tag{5.2}$$

If x is an element in $\mathbf{M}_q$ with nonzero norm, we can consider in $\mathbf{M}_q$

$$\frac{\bar{x}}{n(x)} \tag{5.3}$$

and verify that it is a solution to the equation $x \cdot y = 1$ in the variable y; moreover, since (5.3) possesses the formal properties of the inverse, we can call it inverse of x and denote it by x^{-1}.

3. Let us set

$$y^2 = \frac{1}{\varepsilon}q(\mathbf{x}) \qquad \text{and so} \qquad n(x) = x_0^2 - \varepsilon y^2$$

where y and ε belong to $\mathbf{F}$ or to one of its extensions $\mathbf{F}^o$; in the sequel, we shall consider $\mathbf{M}_q$ on $\mathbf{F}^o$ and we shall exclude the case y identically equal to zero.

[1] A. A. Albert, *Quadratic forms permitting composition*, Ann. of Math., **43** (1942), 161–177.

[2] C. C. Chevalley, *The algebraic theory of spinors*, New York 1954, Chapter 11, § 1.

We will say that a function $w(x)$ in $\mathbf{M}_q$ is *biholomorphic* if

$$w(x) = u(x_0, y) + \frac{1}{y}v(x_0, y)\mathbf{x} \tag{5.4}$$

where $u(x_0, y)$ and $v(x_0, y)$ are functions of x_0 and y [3] satisfying

$$\frac{\partial u}{\partial x_0} = \frac{\partial v}{\partial y} \qquad \frac{\partial u}{\partial y} = \varepsilon \frac{\partial v}{\partial x_0}. \quad [4] \tag{5.5}$$

Taking into account that

$$(m - 2k)\binom{m}{2k} = (2k + 1)\binom{m}{2k + 1},$$

it is easy to verify that powers of a biholomorphic function

$$w^m = \left(\sum_{k=0}^{[m/2]} (\varepsilon)^k \binom{m}{2k} u^{m-2k} v^{2k} \right) + \frac{1}{y} \left(\sum_{k=0}^{[m/2]} (\varepsilon)^k \binom{m}{2k + 1} u^{m-2k-1} v^{2k+1} \right) \mathbf{x}$$

($[m/2]$ is the integer part of $m/2$) are still biholomorphic functions. Since x and x^{-1} are evidently biholomorphic, it turns out that all linear combinations with constant coefficients of positive or negative powers of a variable are biholomorphic, and the property extends to series if $\mathbf{F}$ is finite or with evaluation.

4. Let us denote by ∂ the operator $i_1 \dfrac{\partial}{\partial x_1} + \cdots + i_n \dfrac{\partial}{\partial x_n}$ and let

$$q^{-1}(\mathbf{x}) = \sum_{j,k=1}^{n} \alpha_{jk} x_j x_k$$

be the quadratic form inverse of $q(\mathbf{x})$. Let us set

$$\Box w = \frac{\partial^2 w}{\partial x_0^2} - q^{-1}(\partial)w, \tag{5.6}$$

and let us show that, if $w_0 = u_0 + \dfrac{1}{y}v_0\mathbf{x}$ is biholomorphic and n is odd, then:

$$\Box^{(n+1)/2} w_0 = 0. \tag{5.7}$$

---

[3]Note that $x_0 + y$ and $x_0 - y$ are solutions of (5.2). Thus we can presume that for an extension to cubic modules, etc. it will be more convenient to consider the expressions that appear when solving algebraic equations with the Lagrange method.

[4]Obviously, the derivations are meant as representations which have the usual formal properties.

To simplify the computations we set

$$u_s = \frac{\partial u_{s-1}}{\partial y}\frac{1}{y}, \qquad v_s = \frac{\partial v_{s-1}}{\partial y}\frac{1}{y} - \frac{v_{s-1}}{y^2} = \frac{\partial}{\partial y}\frac{v_{s-1}}{y}$$

$$w_s = u_s + \frac{1}{y}v_s\mathbf{x} \qquad (s = 1, 2, \ldots),$$

and we show that u_s, v_s satisfy the relations

$$\frac{\partial u_s}{\partial x_0} = \frac{\partial v_s}{\partial y} + 2s\frac{v_s}{y}, \qquad \frac{\partial u_s}{\partial y} = \varepsilon\frac{\partial v_s}{\partial x_0}. \tag{5.8}$$

For $s = 0$, (5.8) reduce to (5.5). So, let us suppose that (5.8) hold for $s - 1$; then

$$\frac{\partial u_s}{\partial x_0} = \frac{1}{y}\frac{\partial^2 u_{s-1}}{\partial x_0 \partial y} = \frac{1}{y}\frac{\partial}{\partial y}\left[\frac{\partial v_{s-1}}{\partial y} + 2(s-1)\frac{v_{s-1}}{y}\right] =$$

$$= \frac{1}{y}\frac{\partial}{\partial y}\left[y v_s + (2s-1)\frac{v_{s-1}}{y}\right] = \frac{\partial v_s}{\partial y} + 2s\frac{v_s}{y},$$

$$\frac{\partial u_s}{\partial y} = \frac{\partial}{\partial y}\left(\frac{1}{y}\frac{\partial u_{s-1}}{\partial y}\right) = \varepsilon\frac{\partial}{\partial y}\left(\frac{1}{y}\frac{\partial v_{s-1}}{\partial x_0}\right) = \varepsilon\frac{\partial v_s}{\partial x_0},$$

which is what we had to prove.

With simple computations we then find

$$\frac{\partial^2 w_s}{\partial x_0^2} = \frac{\partial^2 u_s}{\partial x_0^2} + \frac{1}{y}\frac{\partial^2 v_s}{\partial x_0^2}\mathbf{x} = \frac{1}{\varepsilon}\left[y\frac{\partial u_{s+1}}{\partial y} + (2s+1)u_{s+1}\right] +$$

$$+ \frac{1}{\varepsilon}\left[\frac{\partial v_{s+1}}{\partial y} + (2s+2)\frac{v_{s+1}}{y}\right]\mathbf{x},$$

$$\frac{\partial w_s}{\partial x_j} = \frac{1}{\varepsilon}u_{s+1}\frac{\partial q}{\partial x_j} + \frac{1}{\varepsilon}\frac{v_{s+1}}{y}\frac{\partial q}{\partial x_j}\mathbf{x} + \frac{v_s}{y}i_j$$

$$\frac{\partial^2 w_s}{\partial x_j \partial x_k} = \frac{1}{\varepsilon^2 y}\frac{\partial u_{s+1}}{\partial y}\frac{\partial q}{\partial x_j}\frac{\partial q}{\partial x_k} + \frac{1}{\varepsilon}u_{s+1}a_{jk} - \frac{1}{\varepsilon^2}\frac{v_{s+1}}{y^3}\frac{\partial q}{\partial x_j}\frac{\partial q}{\partial x_k}\mathbf{x} +$$

$$+ \frac{1}{\varepsilon^2 y^2}\frac{\partial v_{s+1}}{\partial y}\frac{\partial q}{\partial x_j}\frac{\partial q}{\partial x_k}\mathbf{x} + \frac{1}{\varepsilon}\frac{v_{s+1}}{y}\left[a_{jk}\mathbf{x} + \frac{\partial q}{\partial x_j}i_k + \frac{\partial q}{\partial x_k}i_j\right].$$

[Editors' Note: the second line of the formula above was $+\frac{1}{\varepsilon}\left[\frac{\partial v_{s+1}}{\partial y} + 2s\frac{v_{s+1}}{y}\right]\mathbf{x}$ in the original text].

Finally

$$\sum_{j,k=1}^{n} \alpha_{jk} \frac{\partial^2 w_s}{\partial x_j \partial x_k} = \frac{1}{\varepsilon} \left[\frac{\partial u_{s+1}}{\partial y} y + n u_{s+1} + \frac{\partial v_{s+1}}{\partial y} \mathbf{x} + (n+1) \frac{v_{s+1}}{y} \mathbf{x} \right]$$

so that, taking into account (5.6),

$$\varepsilon \Box w_s = -(n - 2s - 1) w_{s+1}. \tag{5.9}$$

[Editors' Note: in the original text at the right hand side there was $(n-2s-1)w_{s+1}$.]
Thus, if n is odd, for $s = (n-1)/2$ one has

$$\varepsilon \Box w_{(n-1)/2} = 0, \tag{5.10}$$

namely (5.7).

5. A function w in $\mathbf{M}_q$ will be called JB-*monogenic* (Jordan B-monogenic) if, for $B = \|b_{jk}\|$, $b_{jk} = b_{kj}$, $|B| \neq 0$ ($j, k = 1, \ldots, n$ Editors' note j was i in the original manuscript; recall also that $\|b_{jk}\|$ denotes the matrix with entries b_{ij}), one has

$$\frac{\partial w}{\partial x_0} - \frac{1}{2\varepsilon} \sum_{j,k=1}^{n} b_{jk} \left[\frac{\partial w}{\partial x_j} i_k + i_k \frac{\partial w}{\partial x_j} \right] = 0.$$

As w_0 is biholomorphic, we set

$$S = \frac{1}{2\varepsilon} \sum_{j,k=1}^{n} b_{jk} \frac{\partial q}{\partial x_j} i_k$$

[Editors' Note: in the original text it was $S = \frac{1}{\varepsilon} \sum_{j,k=1}^{n} b_{jk} \frac{\partial q}{\partial x_j} i_k$] and we take into account that

$$2 - \frac{1}{\varepsilon y^2} (S\mathbf{x} + \mathbf{x}S) = \frac{1}{\varepsilon y^2} [\mathbf{x}(\mathbf{x} - S) + (\mathbf{x} - S)\mathbf{x}].$$

Then the condition of JB-monogenicity for w_s can be written as

$$0 = \frac{1}{\varepsilon} u_{s+1}(\mathbf{x} - S) + \frac{\partial v_s}{\partial y} \left[1 - \frac{1}{2\varepsilon y^2} \mathbf{x}S + S\mathbf{x} \right] +$$

$$+ \frac{\partial v_s}{\partial y} \left[2s + \frac{1}{2\varepsilon y^2} (\mathbf{x}S + S\mathbf{x}) \right] - \frac{v_s}{y} \frac{1}{\varepsilon} \sum_{j,k=1}^{n} b_{jk} i_j i_k =$$

[Editors' Note: it was $\frac{1}{2\varepsilon^2 y^2}$ in the original text]

$$= \frac{1}{2\varepsilon} [w_{s+1}(\mathbf{x} - S) + (\mathbf{x} - S)w_{s+1}] + \frac{v_s}{y} (2s + 1) - \frac{v_s}{y} \frac{1}{\varepsilon} \sum_{jk} b_{jk} a_{jk};$$

and this is satisfied if $s = \dfrac{n-1}{2}$ and if

$$\mathbf{x} = S. \tag{5.11}$$

Thus: *if (5.11) holds in* $\mathbf{M}_q$, *the* $(n-1)/2$ *power of* $\square$ *of all biholomorphic functions is JB-monogenic.*

6. If $\mathbf{M}_q$ is an algebra, or if B is a scalar and w is such that the jacobian matrix $\partial \mathbf{w}/\partial \mathbf{x}$ is symmetric, we can consider in $\mathbf{M}_q$ the equation

$$\frac{\partial w}{\partial x_0} - \frac{1}{\varepsilon} \sum_{j,k=1}^{n} b_{jk} i_j \frac{\partial w}{\partial x_k} = D_B w = 0. \quad ^5 \tag{5.12}$$

Functions satisfying (5.12) will be called *B-monogenic on the left* (in a similar way one defines functions *B*-monogenic on the right).

With a computation similar to the one done in n. **5** one can see that, *if (5.11) holds, the* $(n-1)/2$ *power of* $\square$ *of biholomorphic functions is B-monogenic on the left and on the right.*

Moreover if $\mathbf{M}_q$ is alternative, multiplying on the left (5.12) by $\bar{D}_B$, the conjugate operator of D_B, one finds that

$$\bar{D}_B D_B w = \left[\frac{\partial^2}{\partial x_0^2} - \frac{1}{\varepsilon^2} g(\partial) \right] w = 0, \tag{5.13}$$

where $g(\mathbf{x})$ is the quadratic form associated with the matrix BAB_{-1}. Thus, *if B satisfies the relation*

$$BAB_{-1} = \varepsilon^2 A^{-1}, \quad ^6 \tag{5.14}$$

then (5.13) coincides with (5.6) and we can say that: *Functions B-monogenic are solutions of the equation* $\square w = 0$.

---

[5]If $\mathbf{M}_q$ is not an algebra, in order that $D_B w$ is in $\mathbf{M}_q$ it is necessary and sufficient that $\partial \mathbf{w}/\partial \mathbf{x} B$ is symmetric; if x is such that the jacobian determinant is always nonzero, this implies that B must be scalar and $\partial \mathbf{w}/\partial \mathbf{x}$ is symmetric.

[6]It is easy to determine the matrices B, provided that one take sinto account that- since B is symmetric - (5.14) can be written as $(BA)^2 = \varepsilon^2 I$. All these relations become then particularly simple in the case of classical quadratic modules, namely for the modules such that

$$f(x) = -(x_1^2 + \cdots + x_n^2).$$

From this and from the preceding result one may reobtain the result of n. **4** (in the present particular case).

7. Let us assume now that **F** is with evaluation, and that the norm $n(x)$ is a definite quadratic form; then in $\mathbf{M}_q$ there are no zero divisors. In a future work, based on the results proved in the preceding sections, we shall show how the theory of quaternionic functions can be extended to functions in the (alternative) algebra of Cayley numbers [Editors' Note: here probably Sce refers to the paper that he eventually wrote with Dentoni, see Chapter 6]; here we will limit ourselves to some considerations on quadratic modules in associative algebras.

With reasonings nowadays classical, we prove first that for B-monogenic functions there is a bilateral integral theorem. Then in the representative space $\mathbf{M}_q$ one can construct an integral formula of Cauchy type, with kernel $\Box^{(n-1)/2}$. From this fact one derives the possibility to develop in series,..., etc. Based on the penultimate paragraph of n. **6**, we also get in this way properties of of functions satisfying the (elliptic) equation $\Box W = 0$.[7]

Let now $\mathbf{M}_q$ be a quadratic module on the real field, and **A** the smallest associative algebra containing it. The problem to extend to the elements of **A** an integral formula (of Cauchy type), once that it has been found for $\mathbf{M}_q$, is trivial if $\mathbf{M}_q$ is of order 2. On the other hand, the problem is not solvable as soon as the order of $\mathbf{M}_q$ is greater than 4, as it turns out from the classification of Clifford algebras[8] and from some simple considerations on the variety of zero divisors in algebras.[9] Thus it remains to be considered only the case in which $\mathbf{M}_q$ has order 4 (and it is not an algebra); but then **A** is a Clifford algebra of order 8 and one would go back to known results, at least in the classical case.[10]

---

[7]All the researches on these topics rely on R. Fueter, *Die Funktionentheorie der Differential-gleichungen $\Delta u = 0$ und $\Delta\Delta u = 0$ mit vier reellen Variablen*, Comment. Math. Helv., v. 7 pp. 307–330 (1934-35). Among the works of Fueter's school, those which treats topics near to ours are: W. Nef, *Funktionentheorie einer Klasse von hyperbolischen und ultrahyperbolischen Differentialgleichungen zweiter Ordnung*, ibid. vol. 17, pp. 83–107 (1944-45); H. G. Haefeli, *Hypercomplexe Differentiale*, ibid., vol. 20, pp. 382–420 (1947); A. Kriszten, *Elliptische systeme von partiellen Differentialgleichungen mit konstanten Koeffizienten*, ibid. vol. 23, pp. 243–271 (1949).

[8]See C. C. Chevalley cited in [(2)]. The classification of classical Clifford algebras can already be found in the paper E. Study, E. Cartan, *Nombres complexes*, Encycl. Franc., I, 5, n. 36, pp. 463.

[9]See M. Sce, *Sulla varietàǎ dei divisori dello zero nelle algebre*, Rend. Lincei, August 1957

[10]G. B. Rizza, *Funzioni regolari nelle algebre di Clifford*, Rend., Roma, v. 15 pp. 53–79 (1956). The integral formulas established in this work hold in general for algebras which are direct sums of quaternions (but not for Clifford algebras of order greater than 8).

5.1   Comments and Historical Remarks

The starting point is a vector space $\mathbf{M}$ on a field $\mathbf{F}$ with characteristic different from 2 and with basis $1 = i_0, i_1, \ldots, i_n$. Following Chevalley [16], one can construct the tensor algebra

$$\mathbf{T} = \oplus_{j=0}^{\infty} \mathbf{M}^{\otimes j}$$

over $\mathbf{M}$ and then assume that $\mathbf{x} \otimes \mathbf{x} = Q(\mathbf{x})$ where

$$Q(\mathbf{x}) = \sum_{j,k=0}^{n} a_{jk} x_j x_k.$$

Note that below we use the symbol q to denote a quaternion, thus here we use Q to denote the quadratic form, although Sce uses q. Below we write $\mathbf{x}^2$ instead of $\mathbf{x} \otimes \mathbf{x}$ and $\mathbf{xy}$ instead of $\mathbf{x} \otimes \mathbf{y}$ and x denotes $x_0 + \mathbf{x}$, $x_0 \in \mathbb{R}$.

The fact that $x^2 \in \mathbf{M}$ implies that $(x + y)^2 \in \mathbf{M}$ and so $\dfrac{xy + yx}{2} \in \mathbf{M}$, thus $\mathbf{M}$ is closed with respect to the operation

$$(x, y) \mapsto x \cdot y = \frac{xy + yx}{2}. \tag{5.15}$$

This multiplicative structure on $\mathbf{M}$ gives in fact a (commutative) Jordan algebra $\mathbf{M}^+$.

Using Sce's terminology, although for us M_Q would be more appropriate, we give the following:

Definition 5.1 We call quadratic module, and we denote it by $\mathbf{M}_q$ the vector space $\mathbf{M}$ in which the multiplicative structure is given by (5.15).

For the sequel it can be useful to keep in mind the references [1, 16, 55, 61, 63, 71, 74, 77] also quoted in the original paper by Sce.

Remark 5.1 As a special case of the previous construction, we can take $\mathbf{F} = \mathbb{R}$ and we can consider, for example, $\mathbf{M} = \mathbb{R}^{n+1}$ identified with the set of paravectors, that is those $x \in \mathbb{R}_n$ that are of the form $x = x_0 + x_1 e_1 + \ldots + x_n e_n$, for $x_0, x_\ell \in \mathbb{R}$, where $e_0 = 1$, and e_ℓ, $\ell = 1, \ldots, n$ are the imaginary units generating a Clifford algebra over $\mathbb{R}$, i.e., $\mathbf{M}$ is the set of paravectors in $\mathbb{R}_n$. If the imaginary units satisfy a nondegenerate bilinear form $B(\cdot, \cdot)$, as in the case of a Clifford algebra, then we can set $Q(\mathbf{x}) = B(\mathbf{x}, \mathbf{x})$ and the construction above corresponds to the construction of a real universal Clifford algebra over n imaginary units. Note that $\mathbf{M}$ can be of signature (p, s), $p + s = n$ namely p units have positive square and s units have negative square. In this case, there exists a basis $e_1^*, \ldots, e_p^*, \ldots, e_n^*$ in which the bilinear form $B(\cdot, \cdot)$ satisfies

1. $B(e_i^*, e_i^*) = 1, i = 1, \ldots, p$;
2. $B(e_i^*, e_i^*) = -1, i = p + 1, \ldots, n = p + s$;
3. $B(e_i^*, e_j^*) = 0, i \neq j$.

The product defined by (5.15) is the classical product of two paravectors in the Clifford algebra.

The Fueter mapping theorem was proved by Fueter in the Mid Thirthies, see [45], and provides an interesting way to generate Cauchy–Fueter regular functions starting from holomorphic functions. The idea is to start with a holomorphic function

$$f_0(u + iv) = \alpha(u, v) + i\beta(u, v)$$

defined in an open set of the upper half complex plane. Given a nonreal quaternion $q = x_0 + \underline{q}$, we define the function

$$f(q) = \alpha(x_0, |\underline{q}|) + \frac{\underline{q}}{|\underline{q}|}\beta(x_0, |\underline{q}|), \qquad (5.16)$$

which is called the quaternionic valued function induced by f_0. Fueter's theorem can be stated as follows (the result was also surveyed in [43, 77]):

Theorem 5.1 (Fueter [45]) *Let $f_0(z) = \alpha(u, v) + i\beta(u, v)$ be a holomorphic function defined in a domain (open and connected) D in the upper-half complex plane and let*

$$\Omega_D = \{q = x_0 + ix_1 + jx_2 + kx_3 = x_0 + \underline{q} \mid (x_0, |\underline{q}|) \in D\}$$

be the open set induced by D in $\mathbb{H}$ and let $f(q)$ be the quaternionic valued function induced by f_0. Then Δf is both left and right Cauchy–Fueter regular in Ω_D, i.e.,

$$\frac{\partial}{\partial \overline{q}} \Delta f(q) = \Delta f(q) \frac{\partial}{\partial \overline{q}} = 0,$$

where Δ is the Laplacian in the four real variables x_ℓ, $\ell = 0, 1, 2, 3$ and $\frac{\partial}{\partial \overline{q}}$ is the Cauchy–Fueter operator.

Almost 20 years later, Sce extended this result in a very pioneering and general way. In the recent literature, Sce's result is known in the following form (see Theorem 5.2 below):

By applying $\Delta^{(n-1)/2}$ (Δ is the Laplacian in $n + 1$ real variables) to a function induced on the set of paravectors by a holomorphic function, one obtains a monogenic one with values in the real Clifford algebra $\mathbb{R}_n$ over an odd number n of imaginary units.

In the sequel, we will discuss mainly the implications of the Fueter–Sce construction in the Clifford setting, so we fix here the notation. The imaginary units of the Clifford algebra $\mathbb{R}_n$ will be denoted by e_ℓ, $\ell = 1, \ldots, n$, and we set $e_0 = 1$. The paravectors are elements of the Clifford algebra that are of the form

$$x = x_0 + x_1 e_1 + \ldots + x_n e_n, \quad x_\ell \in \mathbb{R}, \quad \ell = 0, \ldots n,$$

x_0 is the real (or scalar) part of x also denoted by $\mathrm{Re}(x)$, the 1-vector part of x is defined by $\underline{x} = x_1 e_1 + \ldots + x_n e_n$, the conjugate of x is denoted by $\bar{x} = x_0 - \underline{x}$, and the Euclidean modulus of x is given by $|x|^2 = x_0^2 + \ldots + x_n^2$. The sphere of 1-vectors with modulus 1, is defined by

$$\mathbb{S} = \{\underline{x} = e_1 x_1 + \ldots + e_n x_n \mid x_1^2 + \ldots + x_n^2 = 1\}.$$

We can state now the Clifford algebra version of Sce's theorem in the version that is commonly known in the recent literature.

Theorem 5.2 (Sce [75]) *Consider the Euclidean space $\mathbb{R}^{n+1}$ whose elements are identified with paravectors $x = x_0 + \underline{x}$.*

Let $f_0(z) = f_0(u + iv) = \alpha(u, v) + i\beta(u, v)$ be a holomorphic function defined in a domain (open and connected) D in the upper-half complex plane and let

$$\Omega_D = \{x = x_0 + \underline{x} \mid (x_0, |\underline{x}|) \in D\}$$

be the open set induced by D in $\mathbb{H}$ and $f(x)$ be the Clifford-valued function induced by f_0. Then the function

$$\check{f}(x) := \Delta^{\frac{n-1}{2}} \left(\alpha(x_0, |\underline{x}|) + \frac{x}{|\underline{x}|} \beta(x_0, |\underline{x}|) \right)$$

is left and right monogenic.

For the sequel, it is convenient to define the following maps:

$$T_{FS1} : \alpha(u, v) + i\beta(u, v) \mapsto \alpha(x_0, |\underline{x}|) + \frac{x}{|\underline{x}|} \beta(x_0, |\underline{x}|) \tag{5.17}$$

$$T_{FS2} : \alpha(x_0, |\underline{x}|) + \frac{x}{|\underline{x}|} \beta(x_0, |\underline{x}|) \mapsto \Delta^{\frac{n-1}{2}} \left(\alpha(x_0, |\underline{x}|) + \frac{x}{|\underline{x}|} \beta(x_0, |\underline{x}|) \right) \tag{5.18}$$

Sce's result requires some remarks, in fact it is broader than Theorem 5.2 from two different points of view: the algebra in which it is proven and the type of functions obtained.

Remark 5.2 As Sce observed, the quadratic module $\mathbf{M}_q$ can be embedded only in algebras of specific form: for example in the Cayley–Dickson algebras, in particular the octonions, and in the particular case of associative algebras, in all Clifford algebras or algebras whose semi-simple part is a Clifford algebra. However, Clifford algebras, of any signature, are only a special case of this construction. A natural question is then to ask what happens in the case one considers a module which is not quadratic, but instead cubic or else. It will be interesting to understand if a "Fueter–Sce mapping theorem" can be constructed in that case, and which operator has to be considered instead of the Laplacian.

Remark 5.3 As we said, Sce's extension of Fueter's result is broader than the one commonly quoted in the literature. In fact, with the above notations, it shows that given a function $f_0(z) = f_0(u + iv) = \alpha(u, v) + i\beta(u, v)$ which is holomorphic or anti-holomorphic, see (4.6), then the function

$$\Delta^{\frac{n-1}{2}} f(x) = \Delta^{\frac{n-1}{2}} \left(u(x_0, |\underline{x}|) + \frac{x}{|\underline{x}|} v(x_0, |\underline{x}|) \right) \tag{5.19}$$

is a JB-monogenic function namely it satisfies

$$\frac{\partial f}{\partial x_0} - \frac{1}{2\varepsilon} \sum_{j,k=1}^{n} b_{jk} \left[\frac{\partial f}{\partial x_j} i_k + i_k \frac{\partial f}{\partial x_j} \right] = 0,$$

where the matrix $B = [b_{jk}]$ is is symmetric and nondegenerate. The proof of this result, in the special case when B is scalar and the jacobian matrix of f is symmetric, gives that f is left and right B-monogenic.

Remark 5.4 The case in which one obtains a monogenic function or function in the kernel of the Dirac operator (in the sense of Clifford analysis) is very special and occurs when $B = I$, I being the identity matrix, and $\mathbf{M}_q$ is the set of paravectors in a Clifford algebra. However, the result is proved for an algebra generated by a module with unit and whose elements satisfy a quadratic equation. And again, according to this quadratic equation, the function f_0 satisfies the Cauchy–Riemann equation or a variation of it, see (5.5). Operators of Cauchy–Fueter type in which there are coefficients b_{jk} such that the matrix $B = [b_{jk}]$ is orthogonal have been considered by Shapiro and Vasilevski in [76].

Remark 5.5 About 40 years after Sce, Qian proved in [67] that the theorem of Sce holds in the case of a Clifford algebra over an even number n of imaginary units, using techniques of Fourier multipliers in the space of distributions in order to deal with the fractional powers of the Laplacian. He showed that also in this case, Sce's construction gives a monogenic function. After this paper there have been a number of generalizations and the interested reader may find more information in the survey [69] but see also the papers [38, 64–66, 73]. Qian also gave an interesting application of Fueter–Sce's theorem, see [68], to prove boundedness of singular integral operators.

Remark 5.6 It is also interesting to note that the language of stem functions and induced functions, later used by Sce in his paper on the octonionic case, see Chap. 6, was also previously used by Cullen and Rinehart, see [42, 70]. Also Sudbery in his paper [78] points out that Cullen used functions of the form (5.16) to define an alternative theory of functions of a quaternion variable. The concepts of stem functions, intrinsic and induced functions are relevant in the theory of functions nowadays called slice hyperholomorphic (also called slice regular when they are quaternionic functions and slice monogenic when they have values in a Clifford algebra). In the theory of slice hyperhomolomorphic functions, the two functions u and v (real-valued in the above discussion) have values in the hypercomplex algebra under consideration. In the language of slice hyperholomorphic functions, functions of the form (5.16) with u, v real are called intrinsic, according to the terminology introduced by Cullen and Rinehart [42, 70].

5.2 The Fueter–Sce Theorem: Function and Spectral Theories

The Fueter–Sce–Qian theorem is one of the most fundamental results in complex and hypercomplex analysis because it shows how to generalize complex analysis to the hypercomplex setting. The fact that the generalization procedure is done in two steps means that there are two function theories in such an extension. When we consider for example quaternionic valued functions, we obtain slice hyperholomorphicity for the quaternions at the first step, and Fueter regular functions at the second step. The other important example is the Clifford algebra valued functions where we obtain slice hyperholomorphicity for Clifford algebra, and monogenic functions, respectively. This fact has important consequences in operator theory, because in both steps of the Fueter–Sce–Qian contruction the two types of hyperholomorphic functions have a Cauchy formula. From the Cauchy formula of slice hyperholomorphic functions one deduces the notion of S-spectrum and, as a consequence, the spectral theory on the S-spectrum, while on the Cauchy formula of Fueter regular functions or monogenic functions one deduces the notion of monogenic spectrum and the related spectral theory. In this section we show how the two function theories are related, how they induce the associated spectral theories and the connections between them.

It is important to observe that quaternionic quantum mechanics was the main motivation to search for the S-spectrum but hypercomplex analysis has given the tools to identify this spectrum. In fact, in 1936 Birkhoff and von Neumann, see [13], showed that quantum mechanics can be formulated over the real, the complex and the quaternionic numbers. Since then, several papers and books treated this topic, however it is interesting, and somewhat surprising, that an appropriate notion of spectrum for quaternionic linear operators was not present in the literature. The way in which the so-called S-spectrum and the S-functional calculus were discovered in

2006 by Colombo and Sabadini is well explained in the introduction of the book [41], where it is shown how hypercomplex analysis methods allow to identify the notion of S-spectrum of a quaternionic linear operator which, from the physical point of view, seemed to be ineffable.

Before the works of the Italian mathematicians on slice hyperholomorphic functions, this function theory was simply seen an intermediate step in the Fueter–Sce–Qian's construction. These functions have various analogies with the theory of functions of one complex variable, but also crucial differences which make them very interesting. Moreover, they opened the way in the understanding of the spectral theories in the quaternionic and in the Clifford settings.

The literature on hyperholomorphic functions and related spectral theories is nowadays very large, so we mention only some monographs and the references therein. For the function theory of slice hyperholomorphic functions the main references are the books [33, 40, 46, 49], while for the spectral theory on the S-spectrum we mention the books [10, 20, 33, 41]. For the more classic quaternionic and monogenic function theory we refer to the books [14, 25, 44, 52, 54, 72], and for the monogenic spectral theory and applications we suggest the interested reader to consult [58].

It is also worthwhile to mention that also Schur analysis has been considered in the slice hyperholomorphic setting, see the book [8] and in the references therein. Schur analysis in the Fueter setting and related topics have been treated, for example, in the papers [2–4].

The Fueter–Sce Mapping Theorem and Function Theories

In the title of this section and below we will often refer to the Fueter–Sce–Qian mapping theorem as to Fueter–Sce mapping theorem because, for the sake of simplicity, the case of the fractional Laplacian considered by Qian will not be treated.

We start by discussing the recent research area of slice hyperholomorphic functions. The construction of Fueter is carried out for functions defined on open sets of the upper half complex plane but it can be generalized to the whole complex plane. Consider a stem function

$$f_0(z) = \alpha(u, v) + i\beta(u, v), \quad z = u + iv$$

defined in a set $D \subseteq \mathbb{C}$, symmetric with respect to the real axis, and set

$$f(x) = f(u + Iv) = \alpha(u, v) + I\beta(u, v), \tag{5.20}$$

where I is an element in the sphere $\mathbb{S}$ of purely imaginary quaternions or 1-vectors in the case of a Clifford algebra and x is either a quaternion or a paravector. This function is well defined if

$$\alpha(u, -v) = \alpha(u, v) \quad \text{and} \quad \beta(u, -v) = -\beta(u, v)$$

namely if α and β are, respectively, even and odd functions in the variable v. Additionally the pair (α, β) satisfies the Cauchy–Riemann system. This fact was already understood by Sce, see Chap. 6, no. 4, but was not taken into account until the work of Qian [68].

The theory of functions of the form (5.20) was somewhat abandoned until 2006 when Gentili and Struppa introduced in [48] the following definition:

Definition 5.2 Let U be an open set in $\mathbb{H}$ and let $f : U \to \mathbb{H}$ be real differentiable. The function f is said to be (left) slice regular or (left) slice hyperholomorphic in U if for every $I \in \mathbb{S}$, its restriction f_I to the complex plane $\mathbb{C}_I = \mathbb{R} + I\mathbb{R}$ passing through origin and containing I and 1 satisfies

$$\overline{\partial}_I f(u + Iv) := \frac{1}{2} \left(\frac{\partial}{\partial u} + I \frac{\partial}{\partial v} \right) f_I(u + Iv) = 0,$$

on $U \cap \mathbb{C}_I$.

Analogously, a function is said to be right slice regular (or right slice hyperholomorphic) in U if

$$(f_I \overline{\partial}_I)(u + Iv) := \frac{1}{2} \left(\frac{\partial}{\partial u} f_I(u + Iv) + \frac{\partial}{\partial v} f_I(u + Iv)I \right) = 0,$$

on $U \cap \mathbb{C}_I$, every $I \in \mathbb{S}$.

Further developments of the theory of slice regular functions were discussed also in [28] and the above definition was extended by Colombo, Sabadini and Struppa, in [27], (see also [21, 29, 30]) to the Clifford algebra setting for functions $f :$ $U \to \mathbb{R}_n$, defined on an open set U contained in $\mathbb{R}^{n+1}$, where $\mathbb{R}_n$ is the Clifford Algebra over n imaginary units. Slice regular functions according to Definition 5.2 and their generalization to the Clifford algebra, called slice monogenic functions, possess good properties on specific open sets that are called axially symmetric slice domains. When it is not necessary to distinguish between the quaternionic case and the Clifford algebra case we call these functions slice hyperholomorphic.

On these domains, slice hyperholomorphic functions satisfy an important formula, called Representation Formula or Structure Formula, which allows to compute the values of the function once that we know its values on a complex plane $\mathbb{C}_I$.

Definition 5.3 Let $U \subseteq \mathbb{H}$ (or $U \subseteq \mathbb{R}^{n+1}$). We say that U is axially symmetric if, for every $u + Iv \in U$, all the elements $u + Jv$ for $J \in \mathbb{S}$ are contained in U. We say that U is a *slice domain* if $U \cap \mathbb{C}_I \neq \emptyset$ and $U \cap \mathbb{R}$ is a domain in $\mathbb{C}_I$ for every $I \in \mathbb{S}$.

The link with functions of the form (5.16) or (5.20) is provided by the Representation Formula or Structure Formula:

Theorem 5.3 *Let $f : U \to \mathbb{R}_n$ be a slice hyperholomorphic function defined on an axially symmetric slice domain $U \subseteq \mathbb{R}^{n+1}$. Let $J \in \mathbb{S}$ and let $x \pm Jy \in U \cap \mathbb{C}_J$. Then the following equality holds for all $x = u + Iv \in U$:*

$$f(u + Iv) = \frac{1}{2}\Big[f(u + Iv) + f(u - Iv) \Big] + I\frac{1}{2}\Big[J[f(u - Iv) - f(u + Iv)] \Big]$$
$$= \frac{1}{2}(1 - IJ)f(u + Iv) + \frac{1}{2}(1 + IJ)f(u - Iv).$$
$$(5.21)$$

Moreover, for all $u + Kv \subseteq U$, $K \in \mathbb{S}$, there exist two functions α, β, independent of I, such that for any $K \in \mathbb{S}$ we have

$$\frac{1}{2}\Big[f(u+Kv)+f(u-Kv) \Big] = \alpha(u, v), \qquad \frac{1}{2}\Big[K[f(u-Kv)-f(u+Kv)] \Big] = \beta(u, v).$$
$$(5.22)$$

As a consequence we immediately have:

Corollary 5.1 *Let $U \subseteq \mathbb{R}^{n+1}$ be an axially symmetric slice domain, let $D \subseteq \mathbb{R}^2$ be such that $u + Iv \in U$ whenever $(u, v) \in D$ and let $f : U \to \mathbb{R}_n$. The function f is slice hyperholomorphic if and only if there exist two differentiable functions $\alpha, \beta : D \subseteq \mathbb{R}^2 \to \mathbb{H}$, satisfying*

$$\alpha(u, v) = \alpha(u, -v), \qquad \beta(u, v) = -\beta(u, -v)$$

and the Cauchy–Riemann system

$$\begin{cases} \partial_u\alpha - \partial_v\beta = 0 \\ \partial_u\beta + \partial_v\alpha = 0, \end{cases} \qquad (5.23)$$

such that

$$f(u + Iv) = \alpha(u, v) + I\beta(u, v). \qquad (5.24)$$

Thus, slice hyperholomorphic functions according to Definition 5.2 or the analogous definition for slice monogenic functions are in fact functions of the form (5.20) only on axially symmetric slice domains. However, if one defines a function to be slice hyperholomorphic if it is of the form (5.20) where α, β satisfy the above condition, one has that these functions are defined on axially symmetric open sets, not necessarily slice domains.

Thus, starting with functions of the form (5.20), called slice functions, has the advantage that they are defined on more general sets, moreover one can weaken the requests on the two functions α, β requiring, e.g. only continuity, or differentiability or to be of class $\mathscr{C}^k$, thus giving rise to the class of continuous or differentiable or $\mathscr{C}^k$ slice functions.

The class of slice functions can be considered over real alternative algebras, as done by Ghiloni and Perotti in [50, 51]. The idea of considering functions with values in an algebra more general than quaternions is the one followed by Sce in the paper translated in this Chapter. Although in his paper α, β are real valued, it is clear that his discussion involving the Laplacian, which is a real operator, extends to α, β with values in an algebra.

It is also possible to define slice hyperholomorphic functions, as functions in the kernel of the first order linear differential operator (introduced in [36])

$$
Gf = \left(|\underline{x}|^2 \frac{\partial}{\partial x_0} + \underline{x} \sum_{j=1}^{n} x_j \frac{\partial}{\partial x_j} \right) f = 0,
$$

where $\underline{x} = x_1 e_1 + \ldots + x_n e_n$. While, another way to introduce slice hyperholo-morphicity, done by Laville and Ramadanoff in the paper [56], is inspired by the Fueter–Sce mapping theorem. They introduce the so called holomorphic Cliffordian functions defined by the differential equation $D \Delta^m f = 0$ over $\mathbb{R}^{2m+1}$, where D is the Dirac operator. Observe that the definition via the global operator G requires less regularity of the functions with respect to the definition in [56].

Here and in the following we will dedicate less attention to monogenic functions because they are very well known since long time. They are functions $f : U \subseteq \mathbb{R}^{n+1} \to \mathbb{R}_n$, with suitable regularity, that are in the kernel of the Dirac operator. Contrary to the monogenic case, slice hyperholomorphic functions can be defined in different ways, as shown above, not always equivalent, and also for this reason they require more comments.

Inversion of the Fueter–Sce–Qian Mapping Theorem

The inverse of the Fueter–Sce–Qian mapping can be obtained in at least two different ways. The first approach that has been introduced in the paper [32] is based on the Cauchy formula of monogenic functions and leads to an integral formula for the inverse Fueter–Sce–Qian mapping. In what follows we give some hints of the solution of the inversion problem because it is interesting to see how a partial differential equation is solved using methods of hypercomplex analysis. A second method to study the inverse of the Fueter–Sce–Qian mapping is based on the Radon and dual Radon transform. We will not presented this method here, but we refer the interested reader to the paper [39] for more details.

The Fueter–Sce–Qian mapping has range in the subset of monogenic functions given by the subclass of those functions which are axially monogenic. In simple words if U is an axially symmetric open set in $\mathbb{R}^{n+1}$ a left axially monogenic function on the open set U is a function of the form

$$F(x) = A(x_0, r) + I B(x_0, r)$$

where $x = x_0 + Ir$, $r = |\underline{x}| \neq 0$, $I = \underline{x}/|\underline{x}|$, and such that the functions $A = A(x_0, r)$ and $B = B(x_0, r)$ satisfy the Vekua's system, i.e.

$$\begin{cases} \partial_{x_0} A(x_0, r) - \partial_r B(x_0, r) = \frac{n-1}{r} B(x_0, r), \\ \partial_{x_0} B(x_0, r) + \partial_r A(x_0, r) = 0. \end{cases}$$

Thus, given an axially monogenic function F, we construct a Fueter–Sce primitive of F, namely a function f such that

$$\Delta^{\frac{n-1}{2}} f(x) = F(x).$$

This problem has been solved in [32] in the case n is odd and in [11] in the case of any $n \in \mathbb{N}$. It is interesting to observe that for the solution of this problem it is enough to construct a Fueter–Sce primitive of suitable functions constructed via the Cauchy kernel for monogenic functions. Precisely, we consider the Cauchy kernel of monogenic functions

$$\mathscr{G}(x) = \frac{1}{A_{n+1}} \frac{\overline{x}}{|x|^{n+1}}, \qquad x \in \mathbb{R}^{n+1} \setminus \{0\}, \tag{5.25}$$

where

$$A_{n+1} = \frac{2\pi^{(n+1)/2}}{\Gamma(\frac{n+1}{2})}.$$

and we define the kernels

$$\mathscr{N}_n^+(x) = \int_{\mathbb{S}} \mathscr{G}(x - J)\, dS(J), \qquad \mathscr{N}_n^-(x) = \int_{\mathbb{S}} \mathscr{G}(x - J)\, J\, dS(J),$$

where $\mathbb{S}$ is the unit $(n - 1)$-dimensional sphere in $\mathbb{R}^{n+1}$, while $dS(J)$ is a scalar element of area of $\mathbb{S}$. The two functions $\mathscr{N}_n^{\pm}(x)$ are axially monogenic and their Fueter–Sce primitives, obviously not unique, can be obtained as the monogenic extension of the two functions:

$$\mathscr{W}_n^+(x_0) := \frac{\mathscr{C}_n}{\mathscr{K}_n}\, D^{-(n-1)} \frac{x_0}{(x_0^2 + 1)^{(n+1)/2}},$$

$$\mathscr{W}_n^-(x_0) := -\frac{\mathscr{C}_n}{\mathscr{K}_n}\, D^{-(n-1)} \frac{1}{(x_0^2 + 1)^{(n+1)/2}},$$

where the symbol $D^{-(n-1)}$ stands for the $(n-1)$ integrations with respect to x_0 and $\mathscr{C}_n$ and $\mathscr{K}_n$ are given constant that can be calculated explicitly. Then we used an extension lemma based on properties of the solutions of the Dirac equation, so the Fueter–Sce primitives $\mathscr{W}_n^{\pm}(x)$ are obtained by $\mathscr{W}_n^{\pm}(x_0)$ replacing x_0 by $x = x_0 + x_1 e_1 + \dots x_n e_n$. For example, in the case $n = 3$, we have

$$\mathscr{W}_3^+(x) = \frac{1}{2\pi} \arctan x, \qquad \mathscr{W}_3^-(x) = -\frac{1}{2\pi} x \arctan x.$$

So we can state the inverse Fueter–Sce mapping theorem:

Theorem 5.4 *Let us consider an axially monogenic function*

$$F(x) = A(x_0, r) + J B(x_0, r)$$

defined on an axially symmetric domain $U \subseteq \mathbb{R}^{n+1}$. Let Γ be the boundary of an open bounded subset $\mathscr{V}$ of the half plane $\mathbb{R} + J\mathbb{R}^+$ and let

$$V = \{x = u + Jv, \ (u, v) \in \mathscr{V}, \ J \in \mathbb{S}\} \subset U.$$

Moreover suppose that Γ is a regular curve whose parametric equations $y_0 = y_0(s)$, $\rho = \rho(s)$ are expressed in terms of the arc-length $s \in [0, L]$, $L > 0$. Then, the function

$$f(x) = \int_\Gamma \mathscr{W}_n^- \left(\frac{1}{\rho}(x - y_0)\right) \rho^{n-2} (dy_0 \, A(y_0, \rho) - d\rho \, B(y_0, \rho)) \qquad (5.26)$$

$$- \int_\Gamma \mathscr{W}_n^+ \left(\frac{1}{\rho}(x - y_0)\right) \rho^{n-2} (dy_0 B(y_0, \rho) - d\rho A(y_0, \rho))$$

is a Fueter–Sce primitive of $F(x)$ on V, where $\mathscr{W}_n^{\pm}(x)$ are the Fueter–Sce primitive of $\mathscr{N}_n^{\pm}(x)$.

The proof of this result is rather involved and, in the general case, it requires Fourier multipliers in order to give meaning to fractional powers of the Laplacian. As the Fueter–Sce mapping theorem, also its inversion can be proved in various framework. It was proved for axially monogenic functions of degree k in [35] for n odd, and in the general case in [12]. The case of polyaxially monogenic functions seems to be more complicated and, at the moment, only the biaxial case has been considered in [37].

The Fueter–Sce Mapping Theorem and Spectral Theories

One of the most important motivations for the study hyperholomorphic functions theories is that they induce spectral theories through their Cauchy formulas.

In fact, in quaternionic operator theory a precise notion of spectrum for quaternionic linear operator was missing at least since the paper [13] of G. Birkhoff and J. von Neumann, where they proved that quantum mechanic can be formulated also on quaternionic numbers, but from the operator theory point of view the notion of spectrum of quaternionic linear operators was not made precise. In fact, in all the papers dealing with quaternionic quantum mechanic the notion of right eigenvalues is used, but as it is well known, a part from the finite dimensional case, the right eigenvalues alone are insufficient to construct a quaternionic spectral theory.

It was only in 2006 that, using techniques based solely on slice hyperholomorphic functions, the precise notion of spectrum of a quaternionic linear operator was identified. This spectrum was called the S-spectrum and since then the literature in quaternionic spectral theory has rapidly grown, see [41] for more information. Later in 2015 (and published in 2016) it was proved also the spectral theorem for quaternionic normal operators based on the S-spectrum, see [6, 7] and perturbation results of quaternionic normal operators can be found in [15]. Beyond the spectral theorem there are more recent developments in the direction of the characteristic operator functions, see [10] and the theory of spectral operators developed in [47].

The quaternionic Riesz–Dunford functional calculus based on the S-spectrum, called S-functional calculus (see for example [5, 22]), was extended also to the case of n-tuples of noncommuting operators using the notion of S-spectrum and the theory of slice monogenic functions, see [26] and the book [33].

An important extension of the S-functional calculus to unbounded sectorial operators is the H^∞-functional calculus which is one of the ways to define functions of unbounded operators. The H^∞-functional calculus has been used to define fractional powers of quaternionic linear operators that define fractional Fourier laws for nonhomogeneous material in the theory of heat propagation. For the original contributions see [9, 18, 19]. For a systematic and recent treatment of quaternionic spectral theory on the S-spectrum and the fractional diffusion problems based on these techniques, see the books [20, 41]. Moreover, in the monograph [33] one can find also the foundations of the spectral theory on the S-spectrum for n-tuples of noncommuting operators.

Below, we summarize in the following some of the applications and research directions of the hyperholomorphic function theories and relative spectral theories, induced by the two steps of the Fueter–Sce construction.

1. The *first step* generates *slice hyperholomorphic functions* and the *spectral theory of the S-spectrum*. Among the applications we mention:

- The mathematical tools for quaternionic quantum mechanics, related to the Spectral Theorem based on the S-spectrum.
- New classes of fractional diffusion problems that are based on the definition of the fractional powers of vector linear operators.
- The characteristic operator functions and applications to linear system theory.
- Quaternionic spectral operators, which allow to consider a class of nonself-adjoint problems.
- Spectral theory of Dirac operators on manifolds in the nonself-adjoint case.

2. The *second step* in the Fueter–Sce construction generates *Fueter regular or monogenic functions* and the *spectral theory on the monogenic spectrum*, and some of the applications are:

- Boundary value problems treated with quaternionic techniques, see the book of Gürlebeck and Sprössig [53] and the references therein.
- Quaternionic approach to div-rot systems of partial differential equations, see [34].
- Harmonic analysis in higher dimension, see the work of McIntosh, Qian, and many others [57, 59, 60, 62, 68].

For operator theory the most appropriate definition of slice hyperholomorphic functions is the one that comes from the Fueter–Sce mapping theorem because it allows to assume that the functions are defined only on axially symmetric open sets. The definition below generalizes Fueter's construction from open sets in the upper half complex plane to more general open sets.

Definition 5.4 Let $U \subseteq \mathbb{R}^{n+1}$ be an axially symmetric open set and let $\mathcal{U} \subseteq \mathbb{R} \times \mathbb{R}$ be such that $x = u + Jv \in U$ for all $(u, v) \in \mathcal{U}$. We say that a function $f : U \to \mathbb{R}_n$ of the form

$$f(x) = \alpha(u, v) + J\beta(u, v)$$

is left slice hyperholomorphic if α, β are $\mathbb{R}_n$-valued differentiable functions such that

$$\alpha(u, v) = \alpha(u, -v), \quad \beta(u, v) = -\beta(u, -v) \quad \text{for all } (u, v) \in \mathcal{U}$$

and if α and β satisfy the Cauchy–Riemann system

$$\partial_u \alpha - \partial_v \beta = 0, \quad \partial_v \alpha + \partial_u \beta = 0.$$

It is called right slice hyperholomorphic when f is of the form

$$f(x) = \alpha(u, v) + \beta(u, v)J$$

and α, β satisfy the above conditions.

Since we will restrict just to left slice hyperholomorphic function on U we introduce the symbol $SH(U)$ to denote them.

Theorem 5.5 *Let $U \subseteq \mathbb{R}^{n+1}$ be an axially symmetric open set such that $\partial(U \cap \mathbb{C}_I)$ is union of a finite number of continuously differentiable Jordan curves, for every $I \in \mathbb{S}$. Let f be an $\mathbb{R}_n$-valued slice hyperholomorphic function on an open set* *containing* $\overline{U}$ *and, for any* $I \in \mathbb{S}$, *we set* $ds_I = -Ids$. *Then, for every* $x \in U$, *we have:*

$$f(x) = \frac{1}{2\pi} \int_{\partial(U \cap \mathbb{C}_I)} S_L^{-1}(s, x) ds_I f(s), \tag{5.27}$$

where

$$S_L^{-1}(s, x) = -(x^2 - 2\text{Re}(s)x + |s|^2)^{-1}(x - \bar{s}) \tag{5.28}$$

and the value of the integral (5.27) depends neither on U nor on the imaginary unit $I \in \mathbb{S}$.

It turns out that the kernel $S_L^{-1}(s, x)$ is slice hyperholomorphic in x and right slice hyperholomorphic in s for x, s such that $x^2 - 2\text{Re}(s)x + |s|^2 \neq 0$.

Denoting by $\mathcal{O}(D)$ the set of holomorphic functions on D, by $N(\Omega_D)$ the set of induced functions on Ω_D (which turn out to be intrinsic slice hyperholomorphic functions) and by $AM(\Omega_D)$ the set of axially monogenic functions on Ω_D the Fueter–Sce construction can be visualized by the diagram:

$$\mathcal{O}(D) \xrightarrow{\ T_{FS1}\ } N(\Omega_D) \xrightarrow{\ T_{FS2}=\Delta \ \ (or\ T_{FS2}=\Delta^{(n-1)/2})\ } AM(\Omega_D),$$

where T_{FS1} denotes the first operator of the Fueter–Sce construction and T_{FS2} the second one, see (5.17) and (5.18). The Fueter–Sce mapping theorem induces two spectral theories according to the classe of functions that we consider. The Cauchy formula of slice hyperholomorphic functions allows to define the notion of S-spectrum, while the Cauchy formula for monogenic functions induces the notion of monogenic spectrum, as illustrated by the diagram:

$$
\begin{array}{ccc}
SH(U) & \xrightarrow{\ T_{FS2}\ } & AM(U) \\
\textit{Slice Cauchy formula} \downarrow & & \textit{Monogenic Cauchy formula} \downarrow \\
S - spectrum & & monogenic\ spectrum \\
\downarrow & & \downarrow \\
S - functional\ calculus & & monogenic\ functional\ calculus \\
\downarrow & & \downarrow \\
H^\infty - functional\ calculus & & H^\infty - monogenic\ functional\ calculus
\end{array}
$$

In the above diagram we have replaced the set of intrinsic functions N by the larger set of slice hyperholomorphic functions SH. This is clearly possible because the map T_{FS2} is the Laplace operator or its powers.

Let us consider a Banach space V over $\mathbb{R}$ with norm $\| \cdot \|$. It is possible to endow V with an operation of multiplication by elements of $\mathbb{R}_n$ which gives a two-sided module over $\mathbb{R}_n$ and by V_n we indicate the two-sided Banach module over $\mathbb{R}_n$ given by $V \otimes \mathbb{R}_n$. Our aim is to construct a functional calculus for n-tuples of not necessarily commuting operators using slice hyperholomorphic functions. So we consider the so called paravector operator

$$T = T_0 + \sum_{j=1}^{n} e_j T_j,$$

where $T_\mu \in B(V)$ for $\mu = 0, 1, ..., n$, and where $B(V)$ is the space of all bounded $\mathbb{R}$-linear operators acting on V.

The notion of S-spectrum follows from the Cauchy formula of slice hyperholomorphic functions and from some not trivial considerations on the fact that we can replace in the Cauchy kernel $S_L^{-1}(s, x)$ the paravector x by the paravector operator T also in the case the components $(T_0, T_1, ..., T_n)$ of T do not commute among themselves. We have the following definition.

Definition 5.5 (S-Spectrum) Let $T \in B(V_n)$ be a paravector operator. We define the S-spectrum $\sigma_S(T)$ of T as:

$$\sigma_S(T) = \{s \in \mathbb{R}^{n+1} \ : \ T^2 - 2\operatorname{Re}(s)T + |s|^2 I \ \text{ is not invertible in } B(V_n)\}$$

where I denotes the identity operator. It's complement

$$\rho_S(T) = \mathbb{R}^{n+1} \setminus \sigma_S(T)$$

is called the S-resolvent set.

Definition 5.6 Let $T \in B(V_n)$ be a paravector operator and $s \in \rho_S(T)$. We define the left S-resolvent operator as

$$S_L^{-1}(s, T) := -(T^2 - 2\operatorname{Re}(s)T + |s|^2 I)^{-1}(T - \bar{s}I). \tag{5.29}$$

A similar definition can be given for the right resolvent operator.

Definition 5.7 We denote by $SH_{\sigma_S(T)}$ the set of slice hyperholomorphic functions defined on the axially symmetric set U that contains the S-spectrum of T.

A crucial result for the definition of the S-functional calculus is that integral

$$\frac{1}{2\pi} \int_{\partial(U \cap \mathbb{C}_I)} S_L^{-1}(s, T) \, ds_I \, f(s), \quad \text{for} \quad f \in SH_{\sigma_S(T)} \tag{5.30}$$

depends neither on U nor on the imaginary unit $I \in \mathbb{S}$, so the S-functional calculus turns out to be well defined.

Definition 5.8 (*S*-Functional Calculus) Let $T \in B(V_n)$ and let $U \subset \mathbb{H}$ be as above. We set $ds_I = -Ids$ and we define the *S*-functional calculus as

$$f(T) := \frac{1}{2\pi} \int_{\partial(U \cap \mathbb{C}_I)} S_L^{-1}(s, T) \, ds_I \, f(s), \qquad \text{for} \quad f \in SH_{\sigma_S(T)}. \tag{5.31}$$

The definition of the *S*-functional calculus is one of the most important results in noncommutative spectral theory.

Here we will not enter into the details of the monogenic functional calculus, we just point out that the starting point for its definition is the monogenic Cauchy formula and the fact that one has to give meaning to the monogenic Cauchy kernel (5.25)

$$\mathcal{G}(s - x) = \frac{1}{A_{n+1}} \frac{\overline{s - x}}{|s - x|^{n+1}}$$

when we replace the paravector x by the paravector operator T. In this case there are major differences with respect to the slice hyperholomorphic Cauchy kernel when the components $(T_0, T_1, ..., T_n)$ of T do not commute among themselves. Moreover, the operators $T_\mu : V \to V$, $\mu = 1, ...n$, must have real spectrum when considered as linear operators on the real Banach space V and we have to set $T_0 = 0$.

Since we are discussing the consequences of the Fueter–Sce theorem in the next subsection we will show how we can use an integral version of this theorem to define the *F*-functional calculus which is a version of the monogenic functional calculus for *n*-tuples of commuting operators but it is based on the *S*-spectrum.

The Fueter–Sce Theorem in Integral Form and the F-Functional Calculus

The Fueter–Sce mapping theorem in integral form and the *F*-functional calculus where introduced in [31] and further investigated in [17, 23, 24].

We now show how the Fueter–Sce mapping theorem provides an alternative way to define the functional calculus based on monogenic functions. The main idea is to apply the Fueter–Sce operator T_{FS2} to the slice hyperholomorphic Cauchy kernel as illustrated by the diagram:

$$SH(U) \qquad\qquad\qquad AM(U)$$

$$\downarrow \qquad\qquad\qquad\qquad\qquad$$

$$Slice\ Cauchy\ formula \xrightarrow{\ T_{FS2}\ } Fueter\ Sce\ integral\ form$$

$$\downarrow \qquad\qquad\qquad\qquad\quad \downarrow$$

$$S - Functional\ calculus \qquad\quad F - functional\ calculus$$

This procedure generates an integral transform, called the Fueter–Sce mapping theorem in integral form, that allows to define the so called F-functional calculus. This calculus uses slice hyperholomorphic functions and the commutative version of the S-spectrum, but defines a monogenic functional calculus. We just give an idea of how this works. We point out that the operator T_{FS2} has a kernel and one has to pay attention to this fact with the definition of the F-functional calculus, more details are given in [41]. Then, one has to observe that one can apply the powers of Laplacian to both sides of (5.27) obtaining:

$$\Delta^h f(x) = \frac{1}{2\pi} \int_{\partial(U \cap \mathbb{C}_I)} \Delta^h S_L^{-1}(s, x) ds_I f(s)$$

which amounts to compute the powers of the Laplacian applied to the Cauchy kernel $S_L^{-1}(s, x)$. In general, it is not easy to compute $\Delta^h f$ and when we apply Δ^h to the Cauchy kernel written in the form (5.28), we do not get a simple formula. However, $S_L^{-1}(s, x)$ can be written in two equivalent ways as follows.

Proposition 5.1 Let $x, s \in \mathbb{R}^{n+1}$ (or in $\mathbb{H}$ in the quaternionic case) be such that $x^2 - 2x\text{Re}(s) + |s|^2 \neq 0$. Then the following identity holds:

$$\begin{aligned} S_L^{-1}(s, x) &= -(x^2 - 2x\text{Re}(s) + |s|^2)^{-1}(x - \bar{s}) \\ &= (s - \bar{x})(s^2 - 2\text{Re}(x)s + |x|^2)^{-1}. \end{aligned} \tag{5.32}$$

If we use the second expression for the Cauchy kernel we find a very simple expression for $\Delta^h S_L^{-1}(s, x)$.

Theorem 5.6 Let $x, s \in \mathbb{R}^{n+1}$ be such that $x^2 - 2x\text{Re}(s) + |s|^2 \neq 0$. Let

$$S_L^{-1}(s, x) = (s - \bar{x})(s^2 - 2\text{Re}(x)s + |x|^2)^{-1}$$

be the slice monogenic Cauchy kernel and let $\Delta = \sum_{i=0}^n \frac{\partial^2}{\partial x_i^2}$ be the Laplace operator in the variable $x = x_0 + \sum_{i=1}^n x_i e_i$. Then, for $h \geq 1$, we have:

$$\Delta^h S_L^{-1}(s, x) = C_{n,h} (s - \bar{x})(s^2 - 2\text{Re}(x)s + |x|^2)^{-(h+1)}, \tag{5.33}$$

where

$$C_{n,h} := (-1)^h \prod_{\ell=1}^h (2\ell) \prod_{\ell=1}^h (n - (2\ell - 1)).$$

The function $\Delta^h S^{-1}(s, x)$ is slice hyperholomorphic in s for any $h \in \mathbb{N}$ but is monogenic in x only if and only if $h = (n + 1)/2$, namely if and only if h equals the Sce exponent. We define the kernel

$$\mathscr{F}_L(s, x) := \Delta^{\frac{n-1}{2}} S_L^{-1}(s, x)$$

$$= \gamma_n (s - \bar{x})(s^2 - 2\mathrm{Re}(x)s + |x|^2)^{-\frac{n+1}{2}},$$

where

$$\gamma_n := (-1)^{(n-1)/2} 2^{(n-1)/2} (n-1)! \left(\frac{n-1}{2}\right)!$$

which can be used to obtain the Fueter–Sce mapping theorem in integral form.

Theorem 5.7 *Let n be an odd number. Let f be a slice hyperholomorphic function defined in an open set that contains $\overline{U}$, where U is a bounded axially symmetric open set. Suppose that the boundary of $U \cap \mathbb{C}_I$ consists of a finite number of rectifiable Jordan curves for any $I \in \mathbb{S}$. Then, if $x \in U$, the function $\breve{f}(x)$, given by*

$$\breve{f}(x) = \Delta^{\frac{n-1}{2}} f(x)$$

is monogenic and it admits the integral representation

$$\breve{f}(x) = \frac{1}{2\pi} \int_{\partial(U \cap \mathbb{C}_I)} \mathscr{F}_L(s, x) ds_I f(s), \quad ds_I = ds/I, \qquad (5.34)$$

where the integral depends neither on U nor on the imaginary unit $I \in \mathbb{S}$.

Using the Fueter–Sce mapping theorem in integral form (5.34), one can define a functional calculus for monogenic functions $\breve{f} = \Delta^{\frac{n-1}{2}} f$ using slice hyperholomorphic functions and the S-spectrum. The F-functional calculus is based on (5.34) and it is a monogenic functional calculus in the spirit of the functional calculus based on the monogenic spectrum introduced by McIntosh (see the book of B. Jefferies [58]).

In the sequel, we will consider bounded paravector operators T, with commuting components $T_\ell \in B(V)$ for $\ell = 0, 1, \ldots, n$. Such subset of $B(V_n)$ will be denoted by $BC^{0,1}(V_n)$. The F-functional calculus is based on the commutative version of the S-spectrum (often called F-spectrum in the literature). So we define the F-resolvent operators.

Definition 5.9 (F-Resolvent Operators) Let n be an odd number and let $T \in BC^{0,1}(V_n)$. For $s \in \rho_S(T)$ we define the left F-resolvent operator by

$$F_L(s, T) := \gamma_n (sI - \overline{T})(s^2 - (T + \overline{T})s + T\overline{T})^{-\frac{n+1}{2}}, \qquad (5.35)$$

where the operator $\overline{T}$ is defined by

$$\overline{T} = -T_1 e_1 - \cdots - T_n e_n$$

the constants γ_n are given above.

Definition 5.10 (The F-Functional Calculus for Bounded Operators) Let n be an odd number, let $T \in BC^{0,1}(V_n)$ be such that $T = T_1 e_1 + \cdots + T_n e_n$, assume that the operators $T_\ell : V \to V$, $\ell = 1, .., n$ have real spectrum and set $ds_I = ds/I$, for $I \in \mathbb{S}$. Let $SH_{\sigma_S(T)}$ and U be as in Definition 5.7. We define

$$\breve{f}(T) := \frac{1}{2\pi} \int_{\partial(U \cap \mathbb{C}_I)} F_L(s, T) \, ds_I \, f(s). \qquad (5.36)$$

The definition of the F-functional calculus is well posed since the integrals in (5.36) depends neither on U and nor on the imaginary unit $I \in \mathbb{S}$.

References

1. Albert, A.A.: Quadratic forms permitting composition. Ann. Math. **43**, 161–177 (1942)
2. Alpay, D., Shapiro, M.: Reproducing kernel quaternionic Pontryagin spaces. Integr. Equ. Oper. Theory **50**(4), 431–476 (2004)
3. Alpay, D., Shapiro, M., Volok, D.: Rational hyperholomorphic functions in R^4. J. Funct. Anal. **221**(1), 122–149 (2005)
4. Alpay, D., Shapiro, M., Volok, D.: Reproducing kernel spaces of series of Fueter polynomials. In: Operator Theory in Krein spaces and Nonlinear Eigenvalue Problems. Operator Theory: Advances and Applications, vol. 162, pp. 19–45. Birkhäuser, Basel (2006)
5. Alpay, D., Colombo, F., Gantner, J., Sabadini, S.: A new resolvent equation for the S-functional calculus. J. Geom. Anal. **25**(3), 1939–1968 (2015)
6. Alpay, D., Colombo, F., Kimsey, D.P.: The spectral theorem for quaternionic unbounded normal operators based on the S-spectrum. J. Math. Phys. **57**(2), 023503, 27 pp. (2016)
7. Alpay, D., Colombo, F., Kimsey, D.P., Sabadini, I.: The spectral theorem for unitary operators based on the S-spectrum. Milan J. Math. **84**(1), 41–61 (2016)
8. Alpay, D., Colombo, F., Sabadini, I.: Slice hyperholomorphic schur analysis. In: Operator Theory: Advances and Applications, vol. 256, xii+362 pp. Birkhäuser/Springer, Cham (2016).
9. Alpay, D., Colombo, F., Qian, T., Sabadini, I.: The H^∞ functional calculus based on the S-spectrum for quaternionic operators and for n-tuples of noncommuting operators. J. Funct. Anal. **271**(6), 1544–1584 (2016)
10. Alpay, D., Colombo, F., Sabadini, I.: Quaternionic de Branges Spaces and Characteristic Operator Function. SpringerBriefs in Mathematics, Springer, Cham (to appear 2020/2021)
11. Baohua, D., Kou, K.I., Qian, T., Sabadini, I.: On the inversion of Fueter's theorem. J. Geom. Phys. **108**, 102–116 (2016)
12. Baohua, D., Kou, K.I., Qian, T., Sabadini, I.: The inverse Fueter mapping theorem for axially monogenic functions of degree k. J. Math. Anal. Appl. **476**, 819–835 (2019)
13. Birkhoff, G., von Neumann, J.: The logic of quantum mechanics. Ann. Math. **37**, 823–843 (1936)
14. Brackx, F., Delanghe, R., Sommen, F.: Clifford Analysis. Research Notes in Mathematics, vol. 76, x+308 pp. Pitman (Advanced Publishing Program), Boston (1982).
15. Cerejeiras, P., Colombo, F., Kähler, U., Sabadini, I.: Perturbation of normal quaternionic operators. Trans. Am. Math. Soc. **372**(5), 3257–3281 (2019)
16. Chevalley, C.C.: The Algebraic Theory of Spinors. Columbia University Press, New York (1954)
17. Colombo, F., Gantner, J.: Formulations of the F-functional calculus and some consequences. Proc. Roy. Soc. Edinburgh A **146**(3), 509–545 (2016)

18. Colombo, F., Gantner, J.: An application of the S-functional calculus to fractional diffusion processes. Milan J. Math. **86**(2), 225–303 (2018)
19. Colombo, F., Gantner, J.: Fractional powers of quaternionic operators and Kato's formula using slice hyperholomorphicity. Trans. Am. Math. Soc. **370**(2), 1045–1100 (2018)
20. Colombo, F., Gantner, J.: Quaternionic closed operators, fractional powers and fractional diffusion processes. In: Operator Theory: Advances and Applications, vol. 274, viii+322 pp. Birkhäuser/Springer, Cham (2019)
21. Colombo, F., Sabadini, I.: A structure formula for slice monogenic functions and some of its consequences. In: Hypercomplex Analysis. Trends in Mathematics, pp. 101–114. Birkhäuser Verlag, Basel (2009)
22. Colombo, F., Sabadini, I.: On some properties of the quaternionic functional calculus. J. Geom. Anal. **19**(3), 601–627 (2009)
23. Colombo, F., Sabadini, I.: The F-spectrum and the SC-functional calculus. Proc. Roy. Soc. Edinburgh A **142**(3), 479–500 (2012)
24. Colombo, F., Sabadini, I.: The F-functional calculus for unbounded operators. J. Geom. Phys. **86**, 392–407 (2014)
25. Colombo, F., Sabadini, I., Sommen, F., Struppa, D.C.: Analysis of Dirac Systems and Computational Algebra. Progress in Mathematical Physics, vol. 39. Birkhäuser, Boston (2004)
26. Colombo, F., Sabadini, I., Struppa, D.C.: A new functional calculus for noncommuting operators. J. Funct. Anal. **254**(8), 2255–2274 (2008)
27. Colombo, F., Sabadini, I., Struppa, D.C.: Slice monogenic functions. Israel J. Math. **171**, 385–403 (2009)
28. Colombo, F., Gentili, G., Sabadini, I., Struppa, D.C.: Extension results for slice regular functions of a quaternionic variable. Adv. Math. **222**(5), 1793–1808 (2009)
29. Colombo, F., Sabadini, I., Struppa, D.C.: An extension theorem for slice monogenic functions and some of its consequences. Isr. J. Math. **177**, 369–389 (2010)
30. Colombo, F., Sabadini, I., Struppa, D.C.: Duality theorems for slice hyperholomorphic functions. J. Reine Angew. Math. **645**, 85–105 (2010)
31. Colombo, F., Sabadini, I., Sommen, F.: The Fueter mapping theorem in integral form and the F-functional calculus. Math. Methods Appl. Sci. **33**, 2050–2066 (2010)
32. Colombo, F., Sabadini, I., Sommen, F.: The inverse Fueter mapping theorem. Commun. Pure Appl. Anal. **10**, 1165–1181 (2011)
33. Colombo, F., Sabadini, I., Struppa, D.C.: Noncommutative functional calculus. In: Theory and Applications of Slice Hyperholomorphic Functions. Progress in Mathematics, vol. 289, vi+221 pp. Birkhäuser/Springer, Basel (2011)
34. Colombo, F., Luna-Elizarraras, M.E., Sabadini, I., Shapiro, M., Struppa, D.C.: A quaternionic treatment of the inhomogeneous div-rot system. Mosc. Math. J. **12**(1), 37–48, 214 (2012)
35. Colombo, F., Sabadini, I., Sommen, F.: The inverse Fueter mapping theorem using spherical monogenics. Isr. J. Math. **194**, 485–505 (2013)
36. Colombo, F., Gonzalez-Cervantes, J.O., Sabadini, I.: A nonconstant coefficients differential operator associated to slice monogenic functions. Trans. Am. Math. Soc. **365**(1), 303–318 (2013)
37. Colombo, F., Sabadini, I., Sommen, F.: The Fueter primitive of biaxially monogenic functions. Commun. Pure Appl. Anal. **13**, 657–672 (2014)
38. Colombo, F., Pena Pena, D., Sabadini, I., Sommen, F.: A new integral formula for the inverse Fueter mapping theorem. J. Math. Anal. Appl. **417**(1), 112–122 (2014)
39. Colombo, F., Lavicka, R., Sabadini, I., Soucek, V.: The Radon transform between monogenic and generalized slice monogenic functions. Math. Ann. **363**(3–4), 733–752 (2015)
40. Colombo, F., Sabadini, I., Struppa, D.C.: Entire slice regular functions. Springer Briefs in Mathematics, v+118 pp. Springer, Cham (2016)
41. Colombo, F., Gantner, J., Kimsey, D.P.: Spectral theory on the S-spectrum for quaternionic operators. In: Operator Theory: Advances and Applications, vol. 270, ix+356 pp. Birkhäuser/Springer, Cham (2018)

42. Cullen, C.G.: An integral theorem for analytic intrinsic functions on quaternions. Duke Math. J. **32**, 139–148 (1965)
43. Deavours, C.A.: The quaternion calculus. Am. Math. Month. **80**, 995–1008 (1973)
44. Delanghe, R., Sommen, F., Soucek, V.: Clifford Algebra and Spinor-Valued Functions: A Function Theory for the Dirac Operator. Related REDUCE software by F. Brackx and D. Constales. With 1 IBM-PC floppy disk (3.5 inch). Mathematics and its Applications, vol. 53, xviii+485pp. Kluwer, Dordrecht (1992)
45. Fueter, R.: Die Funktionentheorie der Differentialgleichungen $\Delta u = 0$ und $\Delta\Delta u = 0$ mit vier reellen Variablen. Comment. Math. Helv. **7**, 307–330 (1934/1935)
46. Gal, S., Sabadini, I.: Quaternionic Approximation: With Application to Slice Regular Functions. Frontiers in Mathematics, x+221pp. Birkhäuser/Springer, Cham (2019)
47. Gantner, J.: Operator theory on one-sided quaternionic linear spaces: intrinsic S-functional calculus and spectral operators. Mem. Am. Math. Soc. (to appear 2020). arXiv:1803.10524
48. Gentili, G., Struppa, D.C.: A new theory of regular functions of a quaternionic variable. Adv. Math. **216**, 279–301 (2007)
49. Gentili, G., Stoppato, C., Struppa, D.C.: Regular Functions of a Quaternionic Variable. Springer Monographs in Mathematics, x+185 pp. Springer, Heidelberg (2013)
50. Ghiloni, R., Perotti, A.: Slice regular functions on real alternative algebras. Adv. Math. **226**(2), 1662–1691 (2011)
51. Ghiloni, R., Moretti, V., Perotti, A.: Continuous slice functional calculus in quaternionic Hilbert spaces. Rev. Math. Phys. **25**, 1350006, 83 (2013)
52. Gilbert, J.E., Murray, M.A.M.: Clifford Algebras and Dirac Operators in Harmonic Analysis. Cambridge Studies in Advanced Mathematics, vol. 26, viii+334 pp. Cambridge University Press, Cambridge (1991)
53. Gürlebeck, K., Sprössig, W.: Quaternionic Analysis and Elliptic Boundary Value Problems. International Series of Numerical Mathematics, vol. 89, 253pp. Birkhäuser Verlag, Basel (1990)
54. Gürlebeck, K., Habetha, K., Spröig, W.: Application of Holomorphic Functions in Two and Higher Dimensions, xv+390pp. Birkhäuser/Springer, Cham (2016)
55. Haefeli, H.G.: Hypercomplexe differentiale. Comment. Math. Helv. **20**, 382–420 (1947)
56. Laville, G., Ramadanoff, I.: Holomorphic Cliffordian functions. Adv. Appl. Clifford Algebras **8**(2), 323–340 (1998)
57. Li, C., McIntosh, A., Qian, T.: Clifford algebras, Fourier transforms and singular convolution operators on Lipschitz surfaces. Rev. Mat. Iberoamericana **10**, 665–721 (1994)
58. Jefferies, B.: Spectral properties of noncommuting operators. Lecture Notes in Mathematics, vol. 1843. Springer-Verlag, Berlin (2004)
59. Jefferies, B., McIntosh, A.: The Weyl calculus and Clifford analysis. Bull. Aust. Math. Soc. **57**, 329–341 (1998)
60. Jefferies, B., McIntosh, A., Picton-Warlow, J.: The monogenic functional calculus. Studia Math. **136**, 99–119 (1999)
61. Kriszten, A.: Elliptische systeme von partiellen Differentialgleichungen mit konstanten Koeffizienten. Comment. Math. Helv. **23**, 243–271 (1949)
62. McIntosh, A., Pryde, A.: A functional calculus for several commuting operators. Indiana U. Math. J. **36**, 421–439 (1987)
63. Nef, W.: Funktionentheorie einer Klasse von hyperbolischen und ultrahyperbolischen Differentialgleichungen zweiter Ordnung. Comment. Math. Helv. **17**, 83–107 (1944/1945)
64. Pena Pena, D., Sommen, F.: A generalization of Fueter's theorem. Results Math. **49**(3–4), 301–311 (2006)
65. Pena Pena, D., Sommen, F.: Biaxial monogenic functions from Funk-Hecke's formula combined with Fueter's theorem. Math. Nachr. **288**(14–15), 1718–1726 (2015)
66. Pena Pena, D., Sabadini, I., Sommen, F.: Fueter's theorem for monogenic functions in biaxial symmetric domains. Results Math. **72**(4), 1747–1758 (2017)
67. Qian, T.: Generalization of Fueters result to R^{n+1}. Rend. Mat. Acc. Lincei **9**, 111–117 (1997)

68. Qian, T.: Singular integrals on star-shaped Lipschitz surfaces in the quaternionic space. Math. Ann. **310**, 601–630 (1998)
69. Qian, T.: Fueter Mapping Theorem in Hypercomplex Analysis. In: D. Alpay (ed.), Operator Theory, pp. 1491–1507. Springer, Berlin (2015)
70. Rinehart, R.F.: Elements of a theory of intrinsic functions on algebras. Duke Math. J. **27**, 1–19 (1960)
71. Rizza, G.B.: Funzioni regolari nelle algebre di Clifford. Rend. Lincei Roma **15**, 53–79 (1956)
72. Rocha-Chavez, R., Shapiro, M., Sommen, F.: Integral Theorems for Functions and Differential Forms. Chapman & Hall/CRC Research Notes in Mathematics, vol. 428, x+204 pp. Chapman & Hall/CRC, Boca Raton (2002)
73. Sommen, F.: On a generalization of Fueter's theorem. Z. Anal. Anwendungen **19**, 899–902 (2000)
74. Sce, M.: Sulla varietà dei divisori dello zero nelle algebre. Rend. Lincei (1957)
75. Sce, M.: Osservazioni sulle serie di potenze nei moduli quadratici. Atti Accad. Naz. Lincei. Rend. Cl. Sci. Fis. Mat. Nat. **23**, 220–225 (1957)
76. Shapiro, M.V., Vasilevski, N.L.: Quaternionic ψ-hyperholomorphic functions, singular integral operators and boundary value problems. I. ψ-hyperholomorphic function theory. Complex Variables Theory Appl. **27**, 17–46 (1995)
77. Study, E., Cartan, E.: Nombres complexes. Encycl. Franc. I **5**(36), 463pp.
78. Subdery, A.: $^{(*)}$ Quaternionic analysis. Math. Proc. Cambridge Philos. Soc. **85**, 199–225 (1979)

Chapter 6
Regular Functions in the Cayley Algebra

This chapter contains the translation of the paper:

P. Dentoni, M. Sce, *Funzioni regolari nell'algebra di Cayley*, Rend. Sem. Mat. Univ. Padova, **50** (1973), 251–267.

[Editors'Note: in the original text Definition 1, Theorem 1, Lemma 1, Proposition 1, etc. are indicated as D_1, T_1, L_1, P_1, etc. Below we use the standard LaTeX environments but putting the original labels into parenthesis, i.e., Definition (D_1), etc.]

1. Some considerations lead to think that a theory of regular functions having the essential properties of the classical theory of one complex variable, can be performed only in division algebras.[1] It is known that, among the real alternative algebras, the only division algebras are the algebra of real numbers **R**, of complex numbers **C**, of the quaternions Q, of Cayley numbers C.[2] [Editors' Note: the reader should pay attention to the difference between **C** used to denote the complex numbers and C to denote the Cayley numbers]. G. C. Moisil, R. Fueter and other Authors have developed since some time a theory over the quaternions based on a generalization of the Cauchy-Riemann condition

$$\frac{\partial f}{\partial x} + i\frac{\partial f}{\partial y} = 0$$

---

[1]For example, according to a result by M. Sce [12], the division algebras are the only associative algebras in which the components of the regular functions satisfy equations of elliptic type.

[2]See for example R. D. Schafer [14], p. 48. The algebras listed above are precisely the *real algebras with composition* (Hurwitz's Theorem). Any of them can be obtained from the preceding one in a way analogous to the one that from the algebra **R** leads to $\mathbf{C} = \mathbf{R} + \mathbf{R}i$ (Cayley-Dickson process). See e.g. N. Jacobson [8].

© The Editor(s) (if applicable) and The Author(s), under exclusive licence to Springer Nature Switzerland AG 2020
F. Colombo et al., *Michele Sce's Works in Hypercomplex Analysis*,
https://doi.org/10.1007/978-3-030-50216-4_6

(*regular functions*). This theory, although lacking in the classical differential part (derivatives, primitives) has obtained a remarkable success for the part related to integral properties (Cauchy-type theorems) and the link with harmonic functions. The goal of this paper is to extend to the Cayley algebra the results obtained by G. C. Moisil and R. Fueter over the quaternions.

The definition of regular function (D_1, n.3) and some results, like for example, *the link between regular functions and harmonic functions* (P_1, n.3) extends to the Cayley algebra without relevant modifications. However, there are often difficulties due to the lack of associativity. This fact leads in a natural way to introduce in C a particular class of functions characterized by the property, trivial in the associative case, that the function cf, $(c \in C)$ is regular on the right (*biregular functions*) (T_1, n.3).

This class of functions appears to be strictly related to other important classes considered in algebras (*primary functions, intrinsic functions*)[3] which contain as a particular case power series with scalar coefficients. In the first place we extend to the algebra C a result of R. F. Rinehart related to quaternions; precisely, in the Cayley algebra primary functions and intrinsic functions turn out to coincide (T_2 n.4). Then we obtain the extension to C of a theorem established by R. Fueter for the quaternions Q; precisely it turns out that *for any intrinsic, analytic function f, the function $\Delta^3 f$ is biregular (T$_3$, n.6*. In particular, *in the Cayley algebra for any convergent series $f(x) = \sum a_n x^n$ ($a_n \in C$) the function $\Delta^3 f$ is right regular* (C_1 n. 6).

The interest for biregular functions mainly reveals in connection with the *integral theorem*. Despite what happens in the quaternionic case, in the Cayley algebra C there is no integral theorem for a pair of functions f, g regular on the right and on the left, respectively. For the validity of the theorem, *it is needed that one of the two functions is biregular* (T_4, n. 7). The integral theorem allows to get an *integral formula of Cauchy type for regular functions in the algebra C* (T_5, n. 8), assuming as a kernel a suitable biregular function.

From the integral formula follow some classical consequences. In particular, *the components of regular functions are of class $\mathscr{C}^\omega$*.

2. The Cayley Algebra
As it is well known, the Cayley algebra on the field of real numbers $\mathbf{R}$ is the set of ordered pairs of quaternions, with the multiplicative law

$$(q_1, q_1') \cdot (q_2, q_2') = (q_1 q_2 - \bar{q}_2' q_1', q_2' q_1 + q_1' \bar{q}_2) \tag{6.1}$$

where $\bar{q}_2$, $\bar{q}_2'$ are quaternions conjugated of q_2, q_2'.[4]

---

[3]For these functions see e.g. R. F. Rinehart [10].

[4]For the essential properties of the Cayley algebra, see e.g. N. Jacobson [8], R. D. Schafer [14].

Denoting by $i_0 = 1, i_1\ i_2, i_3$ the ordinary basis of the skew field Q of quaternions, it is convenient to consider in C the basis formed by the elements

$$u_h = \begin{cases} (i_h, 0) & \text{for } h = 0, \ldots, 3, \\ (0, i_{h-4}) & \text{for } h = 4, \ldots, 7. \end{cases}$$

The element u_0 works as a *neutral element* in C, and in the sequel will be identified with the neutral scalar element 1.

The algebra C turns out to be *non associative*. It is however alternative,[5] that is, the associator $(x, y, z) = (xy)z - x(yz)$ is a trilinear, *alternating* function in the variables $x, y, z \in C$. In particular, the *alternative laws left and right*

$$a(ax) = a^2 x, \qquad (xa)a = xa^2 \tag{6.2}$$

and the *flexible law*

$$a(xa) = (ax)a \tag{6.3}$$

hold.

Considering the generic element $x = \sum_{h=0}^{7} \xi_h u_h$ in C, we then denote by $\bar{x} = \xi_0 u_0 - \xi_1 u_1 - \cdots - \xi_7 u_7$, the conjugated element of x, by $\mathrm{Tr}(x) = x + \bar{x}$ the *trace* of x, and by $|x|^2 = \sum_{h=0}^{7} \xi_h^2 = N(x)$ the *norm* of x.

The conjugation $x \rightsquigarrow \bar{x}$, as in the case of quaternions, is intrinsically characterized by being the only *involutorial antiautomorphism* of C such that for any $x \in C$ one has $x + \bar{x}, x\bar{x} \in \mathbf{R} \cdot 1$.[6]

We then have[7]

$$x\bar{x} = \bar{x}x = |x|^2, \qquad |xy| = |x| \cdot |y| \tag{6.4}$$

thus any element $x \in C$ admits inverse $x^{-1} = \bar{x}/|x|^2$.[8]

The norm N turns out to be a quadratic form positive definite on C. Its associated bilinear form

$$(x, y) = \frac{1}{2}(x\bar{y} + y\bar{x}) \tag{6.5}$$

---

[5]For the fundamental notions on non associative and in particular alternative algebras, see e.g. R. D. Schafer [14].

[6]See e.g. R. D. Schafer [14], p. 45–49.

[7]See e.g. R. D. Schafer [14], p. 45–46.

[8]The uniqueness of the inverse can be proven as in the associative case, making use of the relations $a(\bar{a}x) = (a\bar{a})x$, $(xa)\bar{a} = x(a\bar{a})$, equivalent to (6.2).

is called *scalar product* of the elements x, $y \in C$, and gives to the algebra C the structure of real Hilbert space. With respect to the scalar product (6.5), the chosen basis $u_0, \ldots, u_7$ in C turns out to be *orthonormal*.

From (6.4) and (6.5) it follows immediately that the notions of norm and scalar product are *independent of the choice of a basis* in C.

3. Regular Functions

Let U be an open set in C, and let $f : U \to C$ be a function of class $\mathscr{C}^1$ in U. Denoting by

$$D = u_0 \frac{\partial}{\partial \xi_0} + \cdots + u_7 \frac{\partial}{\partial \xi_7}, \tag{6.6}$$

we introduce the definition:

Definition 6.1 (D_1) A function f of class $\mathscr{C}^1$ in an open set U of C is said left, right regular in U if[9]

$$Df = \sum_{h=0}^{7} u_h \frac{\partial}{\partial \xi_h} = 0, \qquad fD = \sum_{h=0}^{7} \frac{\partial}{\partial \xi_h} u_h = 0. \tag{6.7}$$

hold, respectively.

The definition D_1 depends on the chosen basis $u_0, \ldots, u_7$ in C. However, *the set of functions regular on the right, left remains the same if in* (6.7) *instead of* $\{u_h\}$ *one chooses another arbitrary orthonormal basis* $\{v_h\}$ *of* C (n. 2.) In fact, the matrix of the change of basis from $\{u_h\}$ to $\{v_h\}$ is orthogonal. The assertion easily follows.[10]

Then, considering the *conjugated operator* of D

$$\bar{D} = u_0 \frac{\partial}{\partial \xi_0} - \cdots - u_7 \frac{\partial}{\partial \xi_7},$$

the relations

$$D\bar{D} = \bar{D}D = \Delta \tag{6.8}$$

hold, with Δ the Laplacian in eight variables. It follows:

Proposition 6.1 (P_1) *Every function* f *right or left regular in an open set* U *of* C *is harmonic, namely:*

$$\Delta f = 0, \qquad in \ U.$$

---

[9] About regular functions in an associative algebra, see e.g. G. C. Moisil [9], R. Fueter [4, 6]. An ample bibliography is in V. Iftimie [7].

[10] See M. Sce [11], p. 32, note (8). See also P. Dentoni [3].

For the proof, it suffices to use (6.8), bearing in mind that, as we shall see at n. **8**, regular functions are of class $\mathscr{C}^{\infty}$. A large class of functions regular on the right and on the left can be constructed using the proposition:

Proposition 6.2 (P_2) *For every scalar valued function α harmonic in U, the function $f = \bar{D}\alpha = \alpha\bar{D}$ is regular on the right and on the left in U.*

The assertion follows immediately from (6.8).

Based on Proposition P_2, regular functions obtained starting from a harmonic scalar function α are called *biregular*. They are characterized by the following properties:

Proposition 6.3 (P_3) *A function $f = \varphi_0 u_0 + \cdots + \varphi_7 u_7$ of class $\mathscr{C}^1$ in an open set U of C is biregular in U if and only if the following relations are satisfied*

$$\frac{\partial \varphi_0}{\partial \xi_0} = \frac{\partial \varphi_1}{\partial \xi_1} + \cdots + \frac{\partial \varphi_7}{\partial \xi_7}; \quad \frac{\partial \varphi_0}{\partial \xi_h} + \frac{\partial \varphi_h}{\partial \xi_0} = 0 \quad (h = 1, \ldots, 7), \tag{6.9}$$

$$\frac{\partial \varphi_h}{\partial \xi_k} = \frac{\partial \varphi_k}{\partial \xi_h}, \quad (h, k = 1, \ldots, 7). \tag{6.10}$$

In fact, (6.9) and (6.10) are necessary and sufficient for the existence of a scalar function α such that

$$\Delta\alpha = 0; \quad \frac{\partial \alpha}{\partial \xi_0} = \varphi_0, \quad \frac{\partial \alpha}{\partial \xi_1} = -\varphi_1, \quad \cdots \quad \frac{\partial \alpha}{\partial \xi_7} = -\varphi_7.$$

Proposition 6.4 (P_4) *A function f regular on the left (right) in an open set U of C is biregular in U if and only if (6.10) holds.*

P_4 follows immediately from the relation

$$Df = \sum_{i,j=0}^{7} \frac{\partial \varphi_j}{\partial \xi_i} u_i u_j = \sum_{1 \le i < j \le 7} \left(\frac{\partial \frac{\partial \varphi_i}{\partial \xi_j} - \varphi_j}{\partial \xi_i} \right) u_i u_j +$$

$$+ \sum_{h=1}^{7} \left(\frac{\partial \varphi_0}{\partial \xi_h} + \frac{\partial \varphi_h}{\partial \xi_0} \right) u_h + \left(\frac{\partial \varphi_0}{\partial \xi_0} - \frac{\partial \varphi_1}{\partial \xi_1} - \cdots - \frac{\partial \varphi_7}{\partial \xi_7} \right) u_0.$$

Another characterization of biregular functions in the algebra C is given by the theorem

Theorem 6.1 (T_1) *A condition necessary and sufficient for a function f, of class $\mathscr{C}^1$ in an open set U of C, to be biregular in U, is that for any element $c \in C$ the function fc is left regular in U.*[11]

---

[11] The theorem can also be stated with reference to right regularity of the function cf.

In fact, let $f = \sum_h \varphi_h u_h$. Keeping in mind that the associator is alternative, one can write

$$D(fc) = \sum_h u_h \left(\frac{\partial f}{\partial \xi_h} c \right) = (Df)c - \sum_h \left(u_h, \frac{\partial f}{\partial \xi_h}, c \right) = \qquad (6.11)$$

$$= (Df)c + \sum_{1 \leq h < k \leq 7} \left(\frac{\partial \varphi_h}{\partial \xi_k} - \frac{\partial \varphi_k}{\partial \xi_h} \right) (u_h, u_k, c).$$

Let now f be biregular in U. By Proposition P_4, the right hand side of (6.11) vanishes, so that fc is left regular. Conversely, if fc is left regular for all c (and so also for $c_0 = u_0$) one has $D(fc) = Df = 0$ and (6.11) reduces to

$$\sum_{1 \leq h < k \leq 7} \left(\frac{\partial \varphi_h}{\partial \xi_k} - \frac{\partial \varphi_k}{\partial \xi_h} \right) (u_h, u_k, c) = 0$$

for all $c \in C$. Taking into account Proposition P_4, to obtain the statement it is enough to prove the lemma

Lemma 6.1 (L_1) *In the algebra C the linear transformations*

$$L_{h,k}(x) = (u_h, u_k, x) \qquad 1 \leq h < k \leq 7$$

are linear independent.

If in the relation

$$\sum_{1 \leq h < k \leq 7} \lambda_{hk} (u_h, u_k, x) = 0 \qquad (\lambda_{hk} \in \mathbf{R}),$$

one sets $x = u_4$, and subsequently $x = u_1$, $x = u_5$, one obtains

$$\lambda_{12} - \lambda_{56} = 0, \qquad \lambda_{56} - \lambda_{47} = 0, \qquad \lambda_{47} + \lambda_{12} = 0,$$

so that $\lambda_{12} = 0$. For the other coefficients λ_{hk} one goes back to the case just considered, by doing a suitable permutation on the elements of a basis.[12]

A remarkable class of functions regular on the right and on the left, which is included in that one of biregular functions, can be obtained starting from *intrinsic functions* in the algebra C. To these functions is devoted n. **4**.

---

[12]Let us consider in C an auxiliary basis of the form $v_0 = u_0$, $v_1 = u_h$, $v_2 = u_k$, $v_3 = u_h u_k$, $v_4 = u_s$, $v_5 = u_s u_h$, $v_6 = u_s u_k$, $v_7 = u_s(u_h u_k)$. One can easily see that the bases $\{u_i\}$, $\{v_i\}$ have the same multiplication tables.

4. Intrinsic Functions in the Cayley Algebra

In an arbitrary algebra A, are called *intrinsic* the functions f which *commute with the algebra automorphisms*, namely such that

$$f(\omega x) = \omega f(x)$$

for any automorphism ω of A.[13] Examples of intrinsic functions are given by *power series* $\sum \alpha_n x^n$, *with scalar coefficients*.

The most important class of intrinsic functions, which comprises the example just mentioned, is made by *primary functions*, which can be seen as obtained by extending ordinary functions of a complex variable $\zeta = \xi + i\eta$ to the algebra. In the Cayley algebra, the definition is the following.[14]

Let

$$\psi(\zeta) = \psi_1(\xi, \eta) + i\psi_2(\xi, \eta)$$

be a function defined in an open set of the complex field $\mathbf{C}$, with values in $\mathbf{C}$, with the conditions

$$\psi_1(\xi, -\eta) = \psi_1(\xi, \eta), \qquad \psi_2(\xi, -\eta) = -\psi_2(\xi, \eta). \tag{6.12}$$

Then, for any generic element x in C, let us consider the canonical decomposition

$$x = \xi_0 \cdot 1 + \hat{x} \tag{6.13}$$

where $\xi_0 = \frac{1}{2}\operatorname{Tr} x$ and $\operatorname{Tr} \hat{x} = 0$. By setting

$$\lambda^2 = N(\hat{x}) = \xi_1^2 + \cdots + \xi_7^2,$$

$X = \left(\frac{1}{\lambda}\right)\hat{x}$, (6.13) rewrites as

$$x = \xi_0 \cdot 1 + \lambda X. \tag{6.14}$$

Given the above, the function

$$f(x) = \psi_1(\xi_0, \lambda) \cdot 1 + X\psi_2(\xi_0, \lambda) \tag{6.15}$$

---

[13] For intrinsic functions in algebras, see R. F. Rinehart [10]. For the case of quaternions, see also C. G. Cullen [2].

[14] In an arbitrary algebra A, the extension of the function $\varphi(\zeta)$ is usually defined by means of the Hermite interpolation formula. It can be verified without difficulties, in a way analogous to the case of quaternions (see R. F. Rinehart [10], Theorem 8.1), that in the Cayley algebra this definition coincides with the one in the text.

which, by (6.12), does not depend on the sign chosen for λ, is called *the primary function generated by the function of a complex variable ψ*. Primary functions in the algebra C turn out to be intrinsic.[15] In fact, for any automorphism ω of C

$$\omega x = \xi_0 \cdot 1 + \lambda \omega X$$

and since $|\omega X| = |X| = 1$, $\mathrm{Tr}\, \omega X = \mathrm{Tr}\, X = 0$,[16] the statement follows. In the present case we have the following theorem, extension of a very well known result in the algebra of quaternions:[17]

Theorem 6.2 (T_2) *In the Cayley algebra, intrinsic functions coincide with primary functions.*

N. **5** is devoted to the proof of this result.

5. It is convenient to put beforehand some lemmas on automorphisms of C.

Lemma 6.2 (L_2) *For any pair of orthogonal elements a, b in C with $|a| = |b| = 1$, $\mathrm{Tr}\, a = \mathrm{Tr}\, b = 0$, there exists an automorphism ω of the algebra C such that $\omega a = a$, $\omega b = -b$.*

In fact, let c be an element in C with $|c| = 1$, orthogonal to the subspace generated by 1, a, b, ab. Denoting by $\widetilde{Q}$ the subspace of C generated by the elements 1, a, c, ac and taking into account (6.2), (6.4), and (6.5), one can verify without difficulty that $\widetilde{Q}$ is a subalgebra, isomorphic to the algebra Q of quaternions. The element b, which by hypothesis is orthogonal to 1, a, c is also orthogonal to ac. In fact, by a very well known property of the scalar product in C[18]

$$(ac, b) = (c, \bar{a}b) = -(c, ab) = 0.$$

In conclusion, b is orthogonal to $\widetilde{Q}$. This said, we consider as automorphism ω the linear transformation given by the *reflection* of C into itself in which the elements of $\widetilde{Q}$ are fixed and those of the subspace $\widetilde{Q}^{\perp}$, orthogonal to $\widetilde{Q}$, change their sign.[19]

Lemma 6.3 (L_3) *For any pair of independent elements x, y in C with $|x| = |y| = 1$, $\mathrm{Tr}\, x = \mathrm{Tr}\, y = 0$, there exists an automorphism ω of the algebra C such that $\omega x = x$, $\omega y \neq y$.*

Let us set

$$a = x, \qquad b = \frac{y - (x, y)x}{|y - (x, y)x|}$$

---

[15] The result is well known in an arbitrary associative algebra. See R. F. Rinehart [10, Theorem 4.4.]

[16] See e.g. N. Jacobson [8], p. 65.

[17] See R. F. Rinehart [10], p. 15.

[18] See e.g. N. Jacobson [8], p. 57, (8).

[19] See N. Jacobson [8], p. 66.

(by hypothesis, $y - (x, y)x \neq 0$), so that $|b| = 1$, Tr $b = 0$, $(a, b) = 0$ and there exists an automorphism ω of C such that $\omega x = x$, $\omega b = -b$ from which it follows

$$\omega y = 2(x, y)x - y \neq y.$$

Lemma 6.4 (L_4) *For any pair of elements x, y in C with $|x| = |y|$, Tr $x =$ Tr $y = 0$, there exists an automorphism ω of C such that $\omega x = y$.*

Obviously, we can assume that $x - y \neq 0$. First, we also suppose that $x + y \neq 0$. By setting $a = (x + y)/|x + y|$, $b = (x - y)/|x - y|$ it turns out that $|a| = |b| = 1$, $\omega a = a$, $\omega b = -b$. The assertion follows. If $x + y = 0$, let a be any element in C with $|a| = 1$ and orthogonal to $1, x$. By setting $b = x/|x|$, the assertion follows from Lemma L_2.

Given the above, we turn to the proof of T_2. Let f be any intrinsic function in the algebra C; we have to prove that f is primary. To this end, in analogy with (6.14), we write

$$f(x) = \varphi_0 \cdot 1 + \mu Y \tag{6.16}$$

with $\varphi_0, \mu \in \mathbf{R}$, Tr $Y = 0$, $|Y| = 1$. Denoting by ω an automorphism of c such that $\omega X = X$ (and so $\omega x = x$) it turns out that

$$\varphi_0 \cdot 1 + \mu Y = f(x) = f(\omega x) = \omega f(x) = \varphi_0 \cdot 1 + \mu \omega Y. \tag{6.17}$$

Let now $\mu \neq 0$. Then (6.17) gives $\omega Y = Y$; from Lemma L_3 it follows necessarily the linear dependence of X and Y, from which $Y = \varepsilon X$ with $\varepsilon = \pm 1$. By setting

$$\psi_1(\xi_0, \lambda, X) = \varphi_0; \qquad \psi_2(\xi_0, \lambda, X) = \varepsilon \mu$$

we can write in any case

$$f(x) = \psi_1(\xi_0, \lambda, X) + X \psi_2(\xi_0, \lambda, X) \tag{6.18}$$

and evidently it turns out

$$\psi_1(\xi_0, -\lambda, -X) = \psi_1(\xi_0, \lambda, X); \qquad \psi_2(\xi_0, -\lambda, -X) = -\psi_2(\xi_0, \lambda, X). \tag{6.19}$$

The functions ψ_1, ψ_2 are independent of X. In fact, let X' be any element in C with Tr $X' = 0$, $|X'| = 1$. By L_4 there exists an automorphism ω of C such that $\omega X = X'$ therefore the relations

$$f(\omega x) = \psi_1(\xi_0, \lambda, X') \cdot 1 + X' \psi_2(\xi_0, \lambda, X')$$

$$\omega f(x) = \psi_1(\xi_0, \lambda, X) \cdot 1 + X' \psi_2(\xi_0, \lambda, X).$$

hold. Since, by hypothesis, $f(\omega x) = \omega f(x)$, we obtain

$$\psi_1(\xi_0, \lambda, X') = \psi_1(\xi_0, \lambda, X) = \psi_1(\xi_0, \lambda)$$

$$\psi_2(\xi_0, \lambda, X') = \psi_2(\xi_0, \lambda, X) = \psi_2(\xi_0, \lambda)$$

namely the assertion. In particular, (6.18) and (6.19) are relations of the type (6.15), (6.12), respectively; in other words, the intrinsic function f coincides with the primary function generated by the function of a complex variable $\psi(\xi + i\eta) = \psi_1(\xi, \eta) + i\psi_2(\xi, \eta)$.

Theorem T_2 is then completely proved.

In the sequel, we consider mainly intrinsic functions generated by functions $\psi(\zeta)$ *holomorphic*. In this case, the functions f are called *intrinsic analytic* and they include, in particular, functions defined by convergent series with positive or negative powers of the variable x in C with scalar coefficients

$$f(x) = \sum_{n=-\infty}^{+\infty} \alpha_n x^n \qquad (\alpha_n \in \mathbf{R}). \tag{6.20}$$

In fact, it is not difficult to verify that (6.20) is the primary function generated by the holomorphic function

$$\psi(\zeta) = \sum_{n=-\infty}^{+\infty} \alpha_n \zeta^n.$$

6. Analytic Intrinsic Functions and Biregular Functions

To consider analytic intrinsic functions allows to construct in the Cayley algebra C an important class of functions right and left regular which is contained in the class of biregular functions (n. **3**).

Precisely, denoting by Δ the Laplacian in the variables $\xi_0, \ldots, \xi_7$, it is valid the following theorem, extension of a known result of R. Fueter over the quaternions:[20]

Theorem 6.3 (T_3) *For any intrinsic function f analytic in an open set U of C, the function*

$$g = \Delta^3 f \tag{6.21}$$

is biregular in U. The functions (6.21), with f analytic intrinsic, coincide precisely with the biregular functions obtained from a harmonic function $\alpha(\xi_0, \xi_1, \ldots, \xi_7)$ which depends only on the quantities ξ_0 and $\xi_1^2 + \cdots + \xi_7^2$.

---

[20] See R. Fueter [4], p. 314.

The first assertion contained in T_3 follows immediately by a more general result by M. Sce, related to quadratic modules.[21]

To show the second part of T_3, we consider first any analytic intrinsic function f. From (6.15) we get without difficulty[22]

$$D\Delta^2 f = -48 \left(\frac{1}{\lambda^3} \frac{\partial^2 \psi_2}{\partial \lambda^2} - 3 \frac{1}{\lambda^4} \frac{\partial \psi_2}{\partial \lambda} + 3 \frac{1}{\lambda^5} \psi_2 \right). \tag{6.22}$$

Thus the function $\alpha = D\Delta^2 f$ is a scalar function. It depends only on the quantities $\xi_0, \lambda^2 = \xi_1^2 + \cdots + \xi_7^2$ since by (6.12) the right hand side of (6.22) is an even function of λ. Moreover (6.8) gives

$$\overline{D}\alpha = \Delta^3 f = g, \qquad \Delta\alpha = Dg = 0$$

so that the biregular function $g = \Delta^3 f$ is obtained by a harmonic function α of the requested type.

Conversely, let now α be a harmonic function of the variables $\xi_0, \xi_1, \ldots, \xi_7$ depending only on the quantities $\xi_0, \lambda^2 = \xi_1^2 + \cdots + \xi_7^2$. First of all, note that α can be seen as a function $\alpha(\xi_0, \lambda)$ of the two real variables ξ_0 and $\lambda = \sqrt{\xi_1^2 + \cdots + \xi_7^2}$, with the property $\alpha(\xi_0, -\lambda) = \alpha(\xi_0, \lambda)$. This said, consider the function

$$\psi_2(\xi_0, \lambda) = \frac{5}{48} \int_0^{\xi_0} \frac{\lambda^3(\xi_0 - \theta) - \lambda(\xi_0 - \theta)^3}{2} \alpha(\theta, 0)d\theta - \tag{6.23}$$

$$- \frac{1}{48} \int_0^{\lambda} \frac{\lambda 3\nu - \lambda\nu^3}{2} \alpha(\xi_0, \nu)d\nu,$$

which, as it can be easily verified, is solution of the equation

$$\frac{1}{\lambda^3} \frac{\partial^2 \psi_2}{\partial \lambda^2} - 3 \frac{1}{\lambda^4} \frac{\partial \psi_2}{\partial \lambda} + 3 \frac{1}{\lambda^5} \psi_2 0 - \frac{1}{48} \alpha. \tag{6.24}$$

Taking into account that, by hypothesis, it holds

$$0 = \frac{\partial^2 \alpha}{\partial \xi_0^2} + \cdots + \frac{\partial^2 \alpha}{\partial \xi_7^2} = \frac{\partial^2 \alpha}{\partial \xi_0^2} + \frac{\partial^2 \alpha}{\partial \lambda^2} + \frac{6}{\lambda} \frac{\partial \alpha}{\partial \lambda},$$

---

[21] See M. Sce [13], p. 224, n. 6. One has to keep in mind also proposition P_4 after observing that f, and so also g, satisfies (6.10).

[22] The calculations can be abbreviated using relations (6.8), (6.9) by M. Sce [13].

it is not difficult to verify that ψ_2 *is a harmonic function of the two variables* ξ_0, λ. Let now

$$\psi_1(\xi_0, \lambda) = \int_0^{\xi_0} \frac{\partial \psi_2}{\partial \lambda}(\theta, 0)d\theta - \int_0^{\lambda} \frac{\partial \psi_2}{\partial \xi_0}(\xi_0, v)dv \tag{6.25}$$

the harmonic function conjugated to ψ_2 and let $\psi(\zeta)$ be the holomorphic function

$$\psi(\xi_0 + i\lambda) = \psi_1(\xi_0, \lambda) + i\psi_2(\xi_0, \lambda).$$

Since $\alpha(\xi_0, \lambda)$ is even in the second argument, from (6.23) and (6.25) it follows that ψ_2 and ψ_1 are odd and even in λ, respectively, so that (6.12) are satisfied. Then let f *be the analytic intrinsic function generated by* ψ. Bearing in mind (6.24), from (6.22) it follows that

$$D\Delta^2 f = \alpha$$

so, recalling (6.8), we can write

$$g = \overline{D}\alpha = \Delta^3 f.$$

Thus g comes from the analytic intrinsic function f through the operator Δ^3. Theorem T_3 is then completely proved.

We point out the following corollary of T_3:

Corollary 6.1 (C_1) *For any power series*

$$f(x) = \sum a_n x^n, \qquad (a_n \in C)[23]$$

converging in an open set U in C, the function

$$g = \Delta^3 f$$

is right regular in U.

For the proof, one first proves as in the classical theorem of derivation term by term, that we can write

$$\left(\sum_n a_n x^n\right)D = \sum_n (a_n x^n)D.$$

Since the function x^n is analytic intrinsic, C_1 follows immediately from T_3, keeping in mind Theorem T_1 of n. **3**.

---

[23]In alternative algebras, the notation ax^n is not ambigous, since the result does not depend on how the single factors are associated. See e.g. R. D. Schafer [14], Theorem 3.1, p. 29.

7. Integral Theorem

Let us now denote by dx^* the *adjoint form*[24] of the differential form $dx = \sum_h d\xi_h u_h$, namely the 7-form

$$dx^* = \sum_h (-1)^{h+1} d\xi_0 \wedge \dots \widehat{d\xi_h} \wedge \dots \wedge d\xi_7 u_h.$$

The integral theorem given by G. C. Moisil and R. Fueter for regular functions in the algebra Q of quaternions,[25] extends to the Cayley algebra in the following form:

Theorem 6.4 (T_4) *Let g be a function of class C^1 in an open set U of C. A necessary and sufficient condition to have*

$$\int_{\Gamma_7} (f\,dx^*)g = 0 \qquad (6.26)$$

for any function f right regular in U and for any 7-cycle Γ_7 of class C^1 homologous to zero in U, is that the function g is biregular in U.[26]

For the proof, let us observe first that, denoted by **d** the exterior differential, we have in U

$$\mathbf{d}((f\,dx^*)g) = \left(\sum_h \left(\frac{\partial f}{\partial \xi_h} u_h \right) g + \sum_h (f u_h) \frac{\partial g}{\partial \xi_h} \right) d\xi_0 \dots d\xi_7.$$

By the Green–Stokes theorem,[27] (6.26) is equivalent to the relation

$$\sum_h (f u_h) \frac{\partial g}{\partial \xi_h} = 0 \qquad (6.27)$$

for any f right regular in U. For $f = 1$ we have in particular that $Dg = 0$, so that we can write

$$\sum_h (f u_h) \frac{\partial g}{\partial \xi_h} = \sum_h \left(f, u_h, \frac{\partial g}{\partial \xi_h} \right)$$

$$= \sum_h \left(u_h, \frac{\partial g}{\partial \xi_h}, f \right)$$

$$= -\sum_h u_h \left(\frac{\partial g}{\partial \xi_h} f \right),$$

---

[24] See e.g. V. Choquel-Bruhat [1], p. 97.

[25] See G. C. Moisil [9], p. 169; R. Fueter [4], p. 312.

[26] A theorem analogous to T_4 holds for the integral $\int_{\Gamma_7} g(dx^* f)$, with f left regular.

[27] See e.g. B. Segre [15], Ch. II, n. 46.

and so (6.27) is equivalent to $D(gc) = 0$ for every $c \in C$. Theorem T_1 gives the assertion.

8. Integral Formula

As shown by Theorem T_4, to obtain in C an integral representation formula for regular functions, analogous to the one of G. C. Moisil and R. Fueter in the quaternions,[28] it is necessary to assume as a kernel a biregular function. The classical kernel $1/x$ is not biregular in the algebra C. However it is an analytic intrinsic function in $C \setminus \{0\}$ (n. **5**), so that the function $\Delta^3(1/x)$ is biregular (Theorem T_3). So we arrive at the theorem:

Theorem 6.5 (T_5) *If f is a right regular function in an open set U in C and Σ_7 is any 7-dimensional surface, closed, of class C^1 contained in U, then*

$$f(z) = \frac{1}{48\pi^4} \int_{\sigma_7} (f(x)dx^*)\Delta^3 \frac{1}{x-z} \tag{6.28}$$

for any z in the interior of Σ_7.[29, 30]

For the proof, by means of Theorem T_4, we reduce to the calculation of the right hand side of (6.28) on a 7-sphere S_7 with center z and a suitable radius r. Taking into account that

$$\Delta^3(1/(x-z)) = -12^2 \cdot 16((x-z)^{-1}/|x-z|^6)$$

and that on S_7 we have $dx^* = -((x-z)/r)d\sigma$, $d\sigma$ being the area element on S_7, we can write [31]

$$\int_{S_7} (f(x)dx^*)\Delta^3 \frac{1}{x-z} = \frac{-12^2 \cdot 16}{r^7} \int_{S_7} f(x)d\sigma.$$

By letting the radius r of S_7 tend to zero and taking into account that the area of S_7 is $\pi^4 r^7/3$ we immediately arrive at (6.28).

In a way similar to the case of regular functions in the algebra of quaternions,[32] from this obtained integral representation we deduce for regular functions in the Cayley algebra the classical consequences. In particular, *the components of regular functions are of class C^ω.*

---

[28] See G. C. Moisil [9], p. 171; R. Fueter [4], p. 318.

[29] An analogous formula holds obviously for left regular functions.

[30] Here we mean that Σ_7 has winding number 1. In the general case, we have the winding number as a factor at the left hand side of (6.28).

[31] Take into account the relation in the note (8).

[32] See R. Fueter [4], p. 319–330; [5], p. 371–378.

References

1. Choquet-Bruhat, V.: Géometrie différentielle et systèmes extérieurs. Dunod, Paris (1968)
2. Cullen, C.G.: An integral theorem for analytic intrinsic functions on quaternions. Duke Math. J. **32**, 139–148 (1965)
3. Dentoni, P.: Funzioni regolari in un'algebra e cambiamenti di base. Rend. Lincei **51**, 274–281 (1971)
4. Fueter, R.: Die Funktionentheorie der Differentialgleichungen $\Delta u = 0$ und $\Delta\Delta u = 0$ mit vier reellen Variablen. Comment. Math. Helv. **7**, 307–330 (1934/1935)
5. Fueter, R.: Über die analytische Darstellung der regulären Funktionen einer Quaternionenvariablen. Comment. Math. Helv. **7**, 307–330 (1934/1935)
6. Fueter, R.: Die Theorie der regularen Funktionen einer quaternionen Variablen. C. R. Congrès Int. Oslo 1936, I, Oslo, 1937, 75–91
7. Iftimie, V.: Fonctions hypercomplexes. Bull. Math. Soc. Sci. Math. R. S. Roumanie **9**, 279–332 (1965)
8. Jacobson, N.: Composition algebras and their automorphims. Rend. Circ. Mat. Palermo **7**, 55–80 (1958)
9. Moisil, G.C.: Sur les quaternions monogènes. Bull. Sci. Math. **LV**, 168–174 (1931)
10. Rinehart, R.F.: Elements of a theory of intrinsic functions on algebras. Duke Math. J. **27**, 1–19 (1960)
11. Sce, M.: Monogeneità e totale derivabilità nelle algebre reali e complesse, I. Rend. Lincei **16**, 30–35 (1954)
12. Sce, M.: Sui sistemi di equazioni a derivate parziali inerenti alle algebre reali. Rend. Lincei **18**, 32–38 (1955)
13. Sce, M.: Sulle serie di potenze nei moduli quadratici. Rend. Lincei **23**, 220–225 (1957)
14. Schafer, R.D.: An Introduction to Nonassociative Algebras. Academic Press, New York (1966)
15. Segre, B.: Forme differenziali e loro integrali. vol. I. Docet, Roma (1951)

6.1 Comments and Historical Remarks

After this paper by Dentoni and Sce, octonions were studied by physicists but not by mathematicians. The paper [12] by K. Nono considers the operator (6.7) to study factorizations of the Laplacian in dimension eight. In [6] X. Li and L. Peng prove, independently, some theorems already proved by Dentoni and Sce. At the end of their paper they put a remark in which they acknowledge that J. Ryan and M. Shapiro pointed out to them during a conference in Beijing that their results were already known, and the paper [2]. And in fact, in the subsequent papers [7] and [8] they quote the results by Dentoni and Sce. One should note that the results in octonionic function theory are somewhat discovered and re-discovered various times, see for example the work [11]. For other developments of the theory see [9], [10]. In our paper [14] we considered three cases of eight dimensional algebras: biquaternions, i.e. $\mathbb{H} \otimes \mathbb{C}$, Clifford algebra over three units $\mathbb{R}_3$ and the octonions $\mathbb{O}$ and the corresponding notions of holomorphicity, see also [1].

Specifically, in the case of biquaternions in which the imaginary units of $\mathbb{H}$ are denoted by $I, J, K = IJ$ ad the imaginary unit of $\mathbb{C}$ is i, we will say that a function $f : \mathbb{BH} \to \mathbb{BH}$ is $\mathscr{D}_q$–regular if $\mathscr{D}_q f = 0$ where

$$\mathscr{D}_q = \frac{\partial}{\partial z_0} + i \left(I \frac{\partial}{\partial z_1} + J \frac{\partial}{\partial z_2} - K \frac{\partial}{\partial z_3} \right),$$

where $z_\ell = x_\ell + i y_\ell$, $\ell = 0, \ldots, 3$.

In the case of $\mathbb{R}_3$ the operator we consider is not the Dirac operator as in classical Clifford analysis, but instead the operator considered by Rizza in [13]. Denoting by e_1, e_2, e_3 the imaginary units of $\mathbb{R}_3$ and by e_A, where $A = \{i_1, \ldots, i_r\}$ is a subset of the power set $\mathscr{P}\{1, 2, 3\}$ of $\{1, 2, 3\}$, the product $e_{i_1} \ldots e_{i_r}$ we say that a function $f : \Omega \subseteq \mathbb{R}_3 \to \mathbb{R}_3$ is ∂_x–regular if

$$\partial_x f = \sum_{A \in \mathscr{P}\{1,2,3\}} e_A \frac{\partial f}{\partial x_A} = 0.$$

To treat the case of the algebra of octonions $\mathbb{O}$, it is convenient to consider $\mathbb{O}$ as the real algebra generated by the basis $\{e_0, e_1, \ldots, e_7\}$ whose units satisfy the relations:

$$e_r e_s = -\delta_{rs} e_0 + \varepsilon_{rst} e_t$$

where δ_{rs} is the Kronecker delta, ε_{rst} are totally antisymmetric in r, s, t and

$$\varepsilon_{rst} = +1 \qquad \text{for} \quad (rst) = (123), (145), (176), (572), (347), (365), (246).$$

An octonion will be denoted by $X = \sum_{r=0}^{7} e_r x_r$, $x_r \in \mathbb{R}$. The operators giving the notion of holomorphicity left and right in this setting are given in (6.7). As discussed by Dentoni and Sce, the operator D appearing in (6.7) factorizes the Laplacian, see (6.8), and a Cauchy formula is proven, see (6.28). In the case of $\mathbb{BH}$ and $\mathbb{R}_3$ the operator is not anymore elliptic, however it is still possible to prove a Cauchy formula. In the case of $\mathbb{BH}$ let us introduce the notations

$$S_3 = \{q_0 + q \; : \; q_0 \in \mathbb{BH}, \; q = x_0 + I x_1 + J x_2 + K x_3 \in \mathbb{H} \; : \; |q| = 1\},$$

$$B_r = \{q_0 + q \; : \; q_0 \in \mathbb{BH}, \; q = x_0 + I x_1 + J x_2 + K x_3 \in \mathbb{H} \; : \; |q| \leq r\}.$$

Note that S_3 is a sphere in $\mathbb{H}$ and that it is a basis for the homology $H_3(\mathbb{BH}\backslash N_{q_0})$, where N_{q_0} is the null–cone with vertex q_0 defined by $N_{q_0} = \{q \in \mathbb{BH}, \, |q - q_0| = 0\}$. The Cauchy formula is given in the next result:

Theorem 6.6 *Let* $f : \mathbb{BH} \to \mathbb{BH}$ *be a function satisfying* $\mathscr{D}_q f = 0$ *in an open set* $D \subset \mathbb{BH}$, $q_0 \in B_r \subset D$ *and let* Σ *be a cycle homological in* $\mathbb{BH}\backslash N_{q_0}$ *to* S_3. *Then*

$$f(q_0) = \frac{1}{2\pi^2} \int_\Sigma G(q, q_0) \, Dq \, f(q)$$

where

$$G(q, q_0) = \frac{\bar{q} - \bar{q}_0}{|q - q_0|^2},$$

and

$$Dq = \sum_{h=0}^{3} e_h dz_1 \wedge \ldots \widehat{dz_h} \wedge \ldots \wedge dz_3.$$

In the case of $\mathbb{R}_3$, we note that $X \in \mathbb{R}_3$ can be written as $X = \omega_1 q_1 + \omega_2 q_2$ where $\omega_1 = (1 + e_{123})/2$, $\omega_1 = (1 - e_{123})/2$ and q_1, q_2 can be identified with quaternions. We then introduce the functions

$$G_i(q_i, p_i) = \frac{\bar{q}_i - \bar{p}_i}{|q_i - p_i|^4} \qquad i = 1, 2$$

and

$$G(q, p) = \sum_{i=1}^{2} \omega_i G_i(q_i, p_i).$$

We have that $G_i(q_i, p_i)$ is the quaternionic Cauchy–Fueter kernel, so that it is regular (both left and right) with respect to q_i in $\mathbb{R}_3 \backslash Z_{i,p_i}$ where

$$Z_{i,p_i} = \{(q_1 - p_1, q_2 - p_2) \, : \, q_j \neq p_j \, j \neq i\},$$

$i = 1, 2$ while $G(q, p)$ is regular in $\mathbb{R}_3 \backslash Z_p$ where $Z_p = \cup_{i=1}^{2} Z_{i,p_i}$.
   We have the following

Theorem 6.7 *Let* $D \subseteq \mathbb{C}_3$ *be an open set and let* Σ_3 *be a hypersurface boundary of a 4–dimensional domain* Δ_4. *Let* $f, g : \mathbb{R}_3 \to \mathbb{R}_3$ *be left and right regular functions respectively. Then*

$$\int_{\Sigma_3} g(\omega_1 Dq_1 + \omega_2 Dq_2) f = \sum_{i=1}^{2} \omega_i \int_{\Delta_4} d_i g_i \wedge Dq_i f_i - g_i Dq_i \wedge d_i f_i$$

where

$$d_i f_i = \sum_{h=0}^{3} \frac{\partial f_i}{\partial x_{jh}} dx_{jh},$$

for $i, j = 1, 2, j \neq i$.

In more recent times, the theory of slice regular functions was extended also to octonions by Gentili and Struppa, see [3] and was then generalized to real alternative algebras by Ghiloni and Perotti. This latter approach links with the theory of stem and primary functions treated by Dentoni and Sce in their paper. The idea of slice regularity over the octonions is based on the fact that also the algebra of octonions can be seen as union of complex planes.

In fact, $\mathbb{S}$ be the unit sphere of purely imaginary octonions, i.e.

$$\mathbb{S} = \{w = \sum_{k=1}^{7} x_k e_k \text{ such that } \sum_{k=1}^{7} x_k^2 = 1\}.$$

Note that if $I \in \mathbb{S}$, then $I^2 = -1$ so the elements of $\mathbb{S}$ behave as imaginary units. Let us denote by $\mathbb{C}_I$ the complex plane whose elements are the complex numbers of the form $x + Iy$, $I \in \mathbb{S}$. We have that

$$\mathbb{O} = \bigcup_{I \in \mathbb{S}} \mathbb{C}_I$$

With this observation at hand, we can give the following:

Definition 6.2 Let Ω be a domain in $\mathbb{O}$. A real differentiable function $f : \Omega \to \mathbb{O}$ is said to be (left) slice regular if, for every $I \in \mathbb{S}$, its restriction f_I to the complex plane $\mathbb{C}_I$ satisfies

$$\bar{\partial}_I f(x + Iy) := \frac{1}{2} (\frac{\partial}{\partial x} + I \frac{\partial}{\partial y}) f_I(x + yI) = 0,$$

on $\Omega \cap \mathbb{C}_I$, for every $I \in \mathbb{S}$.

As in the quaternionic and in the Clifford algebra case, also octonionic functions slice regular in a neighborhood of the origin admits power series expansion, namely, if w denotes the variable, a slice regular function is of the form $f(w) = \sum_{n \geq 0} w^n a_n$, $a_n \in \mathbb{O}$. In particular, when $a_n \in \mathbb{R}$ for all $n \in \mathbb{N}$ these functions are intrinsic in the sense of no. 4 of this chapter.

As proven in [4], slice regular functions in a real alternative algebras and so also over octonions, satisfy the so-called Representation Formula when they are defined on suitable open sets (the axially symmetric s-domains). Thus slice regular functions are in fact of the form

$$f(w) = f(x + Iy) = \alpha(x, y) + I\beta(x, y)$$

where the $\mathbb{O}$-valued functions α and β are, respectively, even and odd in the variable y, and satisfy the Cauchy-Riemann system. In case α and β are real-valued the function f is an example of primary function, in the sense of no. 4 of this chapter.

Remark 6.1 Observing that the Laplacian is a real operator, and applying Theorem (T_3) to a slice regular function f, we conclude that $\Delta^3 f$ is in the kernel of (6.6) thus it is regular in the octonionic sense. This a version of the Fueter-Sce theorem for slice regular functions of a octonionic variable.

It will be very interesting to prove that this theorem admits an inversion, which conjecturally, should be that for functions F octonionic regular of axial type, i.e., of the form $A(x, y) + I B(x, y)$ there exists a slice regular function f such that $\Delta^3 f = F$.

In recent times, it has been proved that there is another way to decompose the space of octonions as union of quaternionic subspaces, see [5]. Taking a triple I, J, K satisfying

$$I, J \in \mathbb{S}, \qquad I \perp J, \qquad K = IJ$$

we can define the row vector $\mathbb{I}$ by

$$\mathbb{I} := (1, I, J, K) \in \mathbb{O}^4.$$

The set of all such row vectors $\mathbb{I}$ is denoted by $\mathcal{N}$ while $\mathbb{H}_\mathbb{I}$ denotes the algebra of quaternions generated by $\mathbb{I}$, i.e., as real vector space

$$\mathbb{H}_\mathbb{I} = \text{span}_\mathbb{R}\{1, I, J, K\}.$$

We can write the octonionic algebra as:

$$\mathbb{O} = \bigcup_{\mathbb{I} \in \mathcal{N}} \mathbb{H}_\mathbb{I}.$$

To give the definition of slice Dirac-regular function the idea is to define the stem functions and then to impose a suitable condition of holomorphy. This is performed in various steps, starting with the notion of intrinsic function:

Definition 6.3 Let $F : \Omega \to \mathbb{O}^4$ be an octonion-valued vector function defined on an open subset Ω of $\mathbb{R}^4$. If F is a $O(3)-$*intrinsic function*, i.e. for any $x \in \Omega$ and for any $g \in O(3)$ such that $gx \in \Omega$, it satisfies

$$F(x) = g^{-1} F(gx), \tag{6.29}$$

then F is called an $\mathbb{O}-$stem function on Ω.

We then have, without giving too many details for which we refer the reader to [5]:

Definition 6.4 Let $[\Omega]$ be an axially symmetric domain in $\mathbb{O}$ and let $f \in C^1([\Omega])$ so that $f(q) = \mathbb{I}F(x)^T$, where $q = \mathbb{I}x^T$, $F = [F_0, F_1, F_2, F_3]$. If F satisfies

$$
\begin{pmatrix}
\partial_{x_0} & -\partial_{x_1} & -\partial_{x_2} & -\partial_{x_3} \\
\partial_{x_1} & \partial_{x_0} & -\partial_{x_3} & \partial_{x_2} \\
\partial_{x_2} & \partial_{x_3} & \partial_{x_0} & -\partial_{x_1} \\
\partial_{x_3} & -\partial_{x_2} & \partial_{x_1} & \partial_{x_0}
\end{pmatrix}
\begin{pmatrix}
F_0 \\
F_1 \\
F_2 \\
F_3
\end{pmatrix}
=
\begin{pmatrix}
0 \\
0 \\
0 \\
0
\end{pmatrix}
\tag{6.30}
$$

then f is called a (left) slice Dirac-regular function in $[\Omega]$.

It is interesting to note that:

Proposition 6.5 Let $[\Omega]$ be an axially symmetric domain in $\mathbb{O}$ and let $f \in S([\Omega]) \cap C^1([\Omega])$. Then f is (left) slice Dirac-regular if and only if

$$
D_{\mathbb{I}} f(q) = 0, \qquad \forall\, q \in [\Omega] \cap \mathbb{H}_{\mathbb{I}} =: \Omega_{\mathbb{I}} \tag{6.31}
$$

and for all $\mathbb{I} \in \mathcal{N}$.

For this class of functions, we can prove various results like an integral representation formula.

References

1. Colombo, F., Sabadini, I., Struppa, D.C.: Dirac equation in the octonionic algebra. Contemp. Math. **251**, 117–134 (2000)
2. Dentoni, P., Sce, M.: Funzioni regolari nell'algebra di Cayley. Rend. Sem. Mat. Univ. Padova **50**, 251–267 (1973)
3. Gentili, G., Struppa, D.C.: Regular functions on the space of Cayley numbers. Rocky Mountain J. Math. **40**, 225–241 (2010)
4. Ghiloni, R., Perotti, A.: Slice regular functions on real alternative algebras. Adv. Math. **226**, 1662–1691 (2011)
5. Jin, M., Ren, G.B., Sabadini, I.: Slice Dirac operator over octonions. Isr. J. Math. (to appear)
6. Li, X., Peng, L.: The Cauchy integral formulas on the octonions. Bull. Belg. Math. Soc. Simon Stevin **9**(1), 47–64 (2002)
7. Li, X., Peng, L.: Three–line theorem on the octonions. Acta Math. Sin. **20**, 483–490 (2004)
8. Li, X., Kai, Z., Peng, L.: Characterization of octonionic analytic functions. Complex Var. Theory Appl. **50**, 1031–1040 (2005)
9. Liao, J.: A substitute for the set of left k-analytic functions is not a right module. Complex Var. Elliptic Equ. **58**, 963–974 (2013)
10. Liao, J.Q., Wang, J.X.: A necessary and sufficient condition for $D(\phi(\epsilon x)) = 0$. Acta Math. Sin. **56**, 597–604 (2013)

11. Lim, S.J., Shon, K.H.: Hyperholomorphic functions and hyper-conjugate harmonic functions of octonion variables. J. Inequal. Appl. **2013**, 77, 8pp. (2013)
12. Nono, K.: On the octonionic linearization of Laplacian and octonionic function theory. Bull. Fukuoka Univ. Ed. III **37**, 1–15 (1988)
13. Rizza, G.B.: Funzioni regolari nelle algebre di Clifford. Rend. di Mat. **XV**, 1–27 (1956)
14. Sabadini, I., Struppa, D.C.: First order differential operators in real dimension eight. Complex Var. Theory Appl. **47**(10), 953–968 (2002)

Index

Printed in the United States
by Baker & Taylor Publisher Services